PRAISE FOR GARY COLLINS

— CABOT ISLAND —

"Collins's focus on an ordinary event taking place under extraordinary circumstances sheds a tender, respectful light on how strength of character can be forged at the anguished intersection of isolation and bereavement."
DOWNHOME MAGAZINE

"The story is intriguing . . ."
THE HALIFAX CHRONICLE HERALD

— THE LAST FAREWELL —

"The writing here is at its best when the danger and beauty of the sea is subtly described."
ATLANTIC BOOKS TODAY

"*The Last Farewell* tells a true story, but Collins's vivid description and well-realized characters make it read like a novel."
THE HALIFAX CHRONICLE HERALD

"Read *The Last Farewell* not only because it is a moving historical tale of needless tragedy but also because it's a book enriched with abundant details of Newfoundland life not so widespread anymore."
THE PILOT

"[*The Last Farewell:*] *The Loss of the Collett* is informative and intriguing, and not merely for experienced sailors or Newfoundlanders."
THE NORTHERN MARINER

PRAISE FOR GARY COLLINS

— SOULIS JOE'S LOST MINE —

"*Soulis Joe's Lost Mine* is a number of stories in one: it's a great mystery-adventure; it's a fascinating look at prospecting for precious metals; and it's a heart-warming story about the importance of family pride."
THE HALIFAX CHRONICLE HERALD

"This tale also serves to cement Collins's status as one of the region's better storytellers; he has a journalist's eye for detail, his writing is crisp and lean and the narrative arc runs smooth and seamless and is well-peppered with shakes of home-spun humour."
ATLANTIC BOOKS TODAY

— WHAT COLOUR IS THE OCEAN? —

"Delightful rhyming story."
RESOURCE LINKS

"Scott Keating's illustrations are an asset to the book. The double-page illustrations revealing the colour of the ocean are particularly successful in conveying the moods of the ocean and the land."
CM: CANADIAN REVIEW OF MATERIALS

"This tale, set by the sea in Newfoundland, is told in a simple repetitive refrain that will capture the imagination of young readers. . . . Illustrations by Scott Keating, award-winning artist and illustrator, capture the beauty of Newfoundland and the many seasons and moods of the ocean."
ATLANTIC BOOKS TODAY

WHERE EAGLES
LIE FALLEN

WHERE EAGLES LIE FALLEN

THE CRASH OF ARROW AIR FLIGHT 1285
GANDER, NEWFOUNDLAND

GARY COLLINS

FLANKER PRESS LIMITED
ST. JOHN'S
2010

Library and Archives Canada Cataloguing in Publication

Collins, Gary, 1949-
    Where eagles lie fallen : the crash of Arrow Air flight 1285, Gander, Newfoundland / Gary Collins.

ISBN 978-1-897317-67-9

    1. Arrow Air Flight 1285 Crash, Gander, N.L., 1985. 2. United States. Army. Airborne Division, 101st--History. 3. Aircraft accidents--Newfoundland and Labrador--Gander. I. Title.

TL553.53.C3C65 2010   363.12'46509718   C2010-904759-1

© 2010 by Gary Collins

ALL RIGHTS RESERVED. No part of the work covered by the copyright hereon may be reproduced or used in any form or by any means—graphic, electronic or mechanical—without the written permission of the publisher. Any request for photocopying, recording, taping or information storage and retrieval systems of any part of this book shall be directed to Access Copyright, The Canadian Copyright Licensing Agency, 1 Yonge Street, Suite 800, Toronto, ON M5E 1E5. This applies to classroom use as well.

PRINTED IN CANADA

Mixed Sources
Cert no. SW-COC-001271
© 1996 FSC
FSC

This text of this book is printed on Ancient Forest Friendly paper, FSC certified, that is chlorine-free and 100% post-consumer waste.

COVER IMAGE: CLINT COLLINS   COVER DESIGN: ADAM FREAKE

—— FLANKER PRESS ——
PO BOX 2522, STATION C ST. JOHN'S, NL, CANADA
TOLL FREE: 1-866-739-4420 WWW.FLANKERPRESS.COM

14 13 12 11 10   2 3 4 5 6 7 8 9

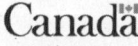   Canada Council   Conseil des Arts
                                 for the Arts   du Canada

We acknowledge the financial support of the Government of Canada through the Book Publishing Industry Development Program (BPIDP) for our publishing activities; the Canada Council for the Arts which last year invested $20.1 million in writing and publishing throughout Canada; the Government of Newfoundland and Labrador, Department of Tourism, Culture and Recreation.

Terrible, sudden death comes with a feeling that sinks into your mind and nauseates your gut. You know that it is true, as unbelievable as it may seem. You know you must face it. You know there is nothing you can do to stop it. You are consumed. Why it happened and how it could be are simply questions. You really don't want to know or even care for the answer. You just want it not to be.

I dedicate this book to the "Fallen Eagles" of the Arrow Air Gander crash. All of them. I guess in the end, war is wherever you find it.

For like a Child sent with a fluttering Light
To feel his way along a gusty Night
Man walks the World: again and yet again
The Lamp shall be by Fits of Passion slain:
But shall not He who sent him from the Door
Relight the Lamp once more, and yet once more?
THE RUBAIYAT OF OMAR KHAYYAM

I carried you on eagles' wings
and brought you to myself.
EXODUS 19:4

INTRODUCTION

I STOOD FOR A long while next to the bronze likeness of the tall soldier guarding the two young children, a small boy and a girl. The soldier bore no arms. The faces of all three wept patina in pale green streaks that always sought out the gleaming metal and quickly aged it, blending it with the ancient land upon which it forever stands. Above me and behind the sculptures, three flags hung limp on their steel masts. The Stars and Stripes of America, the bright red Maple Leaf of Canada, and Newfoundland's own geometric design. A silent witness.

I thought I would find an immediate inspiration here, some hidden muse emanating from the chiselled forms, like you see in the movies or read about in a book. I touched the hand of the boy who carried an olive branch. His metal skin was frigid. His face seemed etched with a faint, hopeful smile. The girl looked up anxiously at her grim-faced protector. The olive branch was missing from her hand. Surely no vandal could stoop so low as to deny the child her peace symbol. My hand felt nothing but the cold of senseless, unfeeling metal. No hidden voices whispering to me.

I sat below the "witnesses" and waited, for what I didn't know. I just somehow knew I should stay for a while where so many had died needlessly.

This cleared track of wooded land above the long Gander Lake reached halfway to the top of a south-facing ridge. It was late April. A few deep, tree-filled valleys still held their cold white blanket of winter. As I pondered what had happened here on this quiet piece of land not all that long ago, the sun burst huge through a hole in the cloud cover and instantly splashed across the soldier and the children's faces. I hadn't noticed before, but this stretch of earth was almost bare of snow, catching every warming angle from the southern sky, while off to the forested sides and lingering among the black tree trunks, trails of dirty snow criss-crossed the ground.

The lake below turned a pleasant, shimmering summer blue with the opening sky. Across the lake, the woody hill rose indifferent. Long, thin orchards of white birch, their branches already reddening, trailed away between the stands of dominant black spruce. On this side of the lake the dark green of the thick softwood stole the colour from the three mute humans, who kept their voiceless vigil at the forest edge.

A faint rustle in the winter-dead grasses at my feet caught my attention. Two American robins pranced in quick, short individual runs that always ended in sudden stops, as with heads askance they looked first at me and then the ground. I hadn't noticed them before.

At this time of year, both the male and the female birds were working side by side, seeking the best building materials for a nearby nest that I couldn't see. I wondered where they had come from. Up from Central America, maybe, following the age-old flyway along with millions of other birds that sought our northern climes every year. Or maybe they had come from the temperate climates of South Carolina or Arkansas, to seek among the spirits of dead countrymen for another round of new life, in a ritual that is as old as time itself.

These small birds, with their splendid colours – red-splashed breast, yellow bill, fringed white wing tips that the talents of human artists could never capture – gave me a part of the answer I was looking for. This place of terrible death wasn't about death at all. It was about continued life. The smallest of winged creatures that had come willingly among the fallen Eagles had shown me.

The robins were not gathering green twigs or sweet-smelling spring bark, or even the sprouting green grasses. They would build their nests from dead yellow grasses and winter-killed dry mosses and long since fallen twigs from which, as always, would come a surety of life anew. These harbingers of every spring already knew what I had searched hard for.

Strange, I thought, maybe there was some hidden voice here. The wonder of it all was not lost on my racing mind. The American robin, probably the most welcome and common spring visitor around our entire northern island, had shown me a way to tell the story of men and women who had taken their namesake from another most majestic of birds, the American Eagle; the white-headed hunter was just as much at home here as it was to the south, in a nation that heralded its beauty and cherished this raptor above all other birds.

Maybe the story of the Screaming Eagles wasn't about their terrible, untimely end at all. It was about the God-given memories they had left behind. A story screaming with its own silence. I wouldn't write about death so much as about the life that goes on and on.

The two robins fluttered away, their beaks too full for even one melody. Silent they flew. Silent the soldier stood with his charge. And just as silent, I crept away.

* * * * *

JUST A FEW DAYS later, I was listening to the local news while driving. An interesting story caught my attention. The reporter said that Lenora Smith, a sixteen-year-old girl in the city of Clarksville, Tennessee, had sent a letter to her mayor and requested that a new bridge being built over the Wilma Rudolph Boulevard in that state be called the Gander Memorial Bridge – in memory of the soldiers of the 101st Airborne Division who perished in the Arrow Air crash in Gander, Newfoundland. The new bridge was to be constructed on the 101st Airborne Division Parkway in Clarksville. It would be a major link between the city and the Army post in Fort Campbell, home of the 101st Airborne.

The teenager, Lenora, had researched the Gander crash for a Veterans Day essay contest in which she was participating. The young girl was steadfast. She drafted two resolutions to the powers that ruled in such matters. It took her two years, but both her resolutions were adopted and passed. The Gander Memorial Bridge, dedicated to the memory of the fallen soldiers, is now an everyday name for that part of America.

I couldn't believe it. This interest coming from one so young – who hadn't even been born when the Arrow Air plane crashed –

filled me with further determination to tell the story of the lives affected by the disaster.

* * * * *

STILL LATER I HEARD about a new ship being christened in the States to commemorate the loss of the USS *Pollux* and USS *Truxtun*. Both ships had been wrecked on Newfoundland's south coast in 1942 with a terrible loss of life. I immediately decided that I should include the sea disaster in the manuscript. It was one of the greatest naval disasters in American history. The story appears in full at the end of this book, as an appendix.

The Arrow Air crash was the greatest one-day loss for the military by air, and the greatest one-day loss of life for the 101st battalion. It took more lives in one day than the brigade suffered at the Siege of Bastogne and even the greater Battle of the Bulge. It was the worst disaster in Canadian aviation history, and also the highest death toll for the American Armed Forces since World War II.

The correlation between the two tragedies was irresistible.

To write about the three ships that ran aground on the south coast of Newfoundland near St. Lawrence, I would first have to visit the location and get a feel for the place.

* * * * *

I WALKED OUT TO Chambers Cove from the town of St. Lawrence, to see where the ships had foundered and where all those men had perished. It was a pleasant traverse over a worn, well-defined gravel trail. The local ATVs had rutted the way, but for the most part the trail was dry and easy walking.

The sign at the path entrance said it would take twenty minutes, so I told my wife, Rose – who agreed to wait in the car because of a problem with her knee – that I'd also need at least twenty minutes once I got out there, as well as the time it would take me to get back. An hour would be good enough.

I made the hike in less than fifteen minutes. The scenery was spectacular. The temperature was great for late May in Newfoundland, around 14°C, although the high wind from the southwest made it very chilly on the headlands. Lining either side of the trail was the stunted growth of evergreens, not much higher than a tall man's head, their slanted tops flattened and bowed by the relentless onshore winds. Newfoundlanders call the short, tangled evergreen trees "tuckamore."

The trail and the posted signs heralding the shipwreck site led me out of the tuckamore growth. I came out on a grassy bluff that spread below me like a greening carpet. There were several low, grey rock fences crafted long ago by skilled hands and bent backs, in an effort to free the sparse soil for farming. The place looked like a deserted Irish croft.

My ears filled with the sound of wind and crashing sea. To my left the blue ocean rolled in the length of the coast and faded in the hazy distance. A few seagulls hovered over the landwash, searching. The tide was out, exposing sodden strings of greenish yellow kelp. There were dozens of tidal pools for the birds to search over. A small, brackish stream had carved a deep yet narrow channel down from the highland, running through a steep pebbled beach to carve a runnel that the high tide would only bury again. From somewhere off to my right and higher up came a steady clamour. I thought the brook might have a hidden waterfall.

I followed the stream up through the narrow gorge and soon realized the sound was coming from higher up the green promontory and away from the stream. There was no waterfall. Standing against the skyline and perched on the very edge of the headland above me were two more signs. I hadn't noticed them before. The noise increased as I climbed the path.

I hurried up the incline – I was pressed for time – and breathing hard soon reached the top. I stood gasping, but not for breath. I looked at the scene below. A magnificent cove shaped like a horseshoe stretched out before me. The sound I had been hearing was the wild wind and wave that bore in from the open sea and the outer edge of the cove. Every approaching roller was capped with a white spray, as if bursting with enthusiasm at the impact it would make on the land it was racing toward.

This was Chambers Cove. This was the place where the ships had run aground, where hundreds of men had died so long ago. My God, what a sight it must have been!

The wind racing up over the steep embankment cut like a razor's edge. It plastered my shirt to my chest and watered my eyes as it roared for freedom up over the lip of the foreland. I was freezing, but I was rooted to the spot. I approached one of the signs on the trail, which turned out to be a remarkable painting.

Boldly depicted there for all to see, with that cruel, magnificent sea as a backdrop, was the terrible long-ago scene. So vivid was the canvas that I thought I could hear the tearing agony of the doomed vessels as they rented themselves upon the cliffs. I wondered if there had been sparks when the forged steel made contact with the ancient rock. Such an impact the painting had on me that day.

Between the roil of the spindrift below and the bottom edge of the headland upon which I stood was a short, relatively dry beach area. There was something else I hadn't noticed before. I couldn't believe my eyes. An American bald eagle was perched on the rocky shoreline!

It was just sitting there inside the surf line, looking out to sea. I didn't believe it even knew I was standing above it. Not only was it an American eagle, but a young one. It wasn't old enough to have yet developed its distinctive "bald" head. New life flourishing around this place made famous for its tale of death.

Was this another omen for me, like the American robin at the Silent Witness Memorial? A crow appeared over the lip of the cliff. It just floated on the invisible updrafts. Then it spotted the young eagle. The eagle saw the diving crow, and looking up noticed me too. Its head cocked to one side as it looked up at the dropping crow. It seemed to ignore me.

I watched as the crow flew into the eagle's space several times, its outstretched black claws always just above the larger bird's gaping beak. The raptor raised its pinions. At first I thought it would fly away, but it was only reminding its tormentor of its dominating size. Why do crows always chase eagles, anyway? It is a common sight around our coast.

The small bird seemed to be toying with the eagle, but soon it tired of its fun and canted upward, looking for more mischief as it took off beyond the cove and out of sight. When I looked down again, the eagle was still there. It was looking up at me, its head tilted. I hurried a few lines into my notebook, wanting to capture my thoughts while they were still fresh. My fingers were numb from the bite of the wind. The ink from my pen was half-frozen, making fragmented lines on the page. I scrawled a few lines about the cove, the seething waves, the cold, loud wind, the forbidding landfall. About the ships. About the 203 young men who had died here. About the painting, the daring crow. About the young eagle.

When I looked down again, the bird was gone, and though I searched the sky for a while, I saw it no more. My head filled with a collision of thoughts that I would probably never figure out, I walked away from Chambers Cove and was almost surprised when the sight of my modern auto brought me into another century.

PROLOGUE

THE SHORTEST DISTANCE ACROSS the broad North Atlantic is between Newfoundland and Ireland. Alcock and Brown made the very first transatlantic flight from Lester's Field, St. John's, on June 14, 1919. Charles Lindbergh made the first solo flight from New York to Paris, flying over Newfoundland on May 20, 1927. He landed in Paris thirty-three and one half hours after leaving Roosevelt Field, New York.

The very first woman to fly alone across the Atlantic, Amelia Earhart, left from a short airfield in Harbour Grace, Newfoundland, on August 20, 1932. When she landed on a bog in northern Ireland fourteen hours, fifty-six minutes later, the thirty-seven-year-old lady from the Kansas plains had written another page of Newfoundland history along with her own.

* * * * *

WITH BUCKSAW AND AXE and sweated manpower and not much more, in the mid-1930s it was cut out of the virgin boreal forest of Newfoundland's hinterland. So remote and secret was the area that only a few caribou hunters and trappers of beaver, fox, and otter had seen this high, wooded plateau that spread itself level above the big blue lake that twisted its way thirty-two miles away to the west. The area – which only a few knew as Hattie's Camp, on the north side of Gander Lake and along by the narrow-gauge island railway – quickly grew in size and out of its name.

The word "war" was on everyone's lips, but none of the workers knew what a major role the place they had cut and fashioned and raised out of the forest would play in the history of the great world battle that was to come. The name Gander would be echoed through countless earpieces on lonely nights and become known as a welcome safe haven from troubled skies for generations to come. From the long lake that heralded the name of the male Canada goose the new settlement took its name, and the brand new airport town of Gander fledged out of the wilderness.

Workers from the island's bays, seeking employment in the lean and hungry last years of the Dirty Thirties, were amazed to find this oasis of activity growing only a day's journey beyond their picket fences. Many of them poled and paddled their long riverboats up the Gander River, fighting the swift current from where it flowed into the ocean at Gander Bay. Reaching the first of many mouths on this huge river, they walked the ten or so miles east through virgin forest, in hopes of obtaining simple labour work. They were seldom disappointed.

All along the Bonavista North shore as well, word trickled through the coastal outports from weekly mailboat visits and from the scattered travellers who walked through the towns on their way home from the railhead at Gambo.

"Lots of work inland, b'ys. They're building something in the woods. Gander, they calls it."

Men from that part of the coast rowed in from their isolated island homes and met with their equally isolated "mainland" friends. All of them, from the Bonavista North area, walked the many forested and trail-twisting miles to the train at Gambo. From here they waited for the slow-moving "Newfie Bullet" to take them to the frontier townsite of Gander.

The island nation of Newfoundland and Labrador had cultivated a strong breed of working men and women unequalled anywhere. The men were seamen, woodsmen, trappers, carpenters, and loggers capable of cutting their own logs for lumber to build houses of their own design and construction, in the most inhospitable of places and on the roughest of grounds. The women were no less adroit. They kept warm, clean, and comfortable homes and managed with lean finances in places where even the staunchest of people might fail. They were also a very loving breed. Their men hated to leave them.

None of the men knew what to expect when they detrained at Gander. All of them were surprised, for here in "the woods," marvellous things were happening. There were many firsts for these men to see at the Gander site. Modern tractors on steel pads moved great plains of earth, feasting on the exposed soil, turning and spreading it to their liking. Sand, dug by hand on the south side of Gander Lake and wheeled in squeaky barrows to waiting flared-nosed scows, was just as laboriously wheeled ashore on the north side and hauled by huge, booming, gas-burning trucks up over the steep ridge to "the site."

Here, great volumes of cement mixed by machine and by hand were blended with the portaged sand to build foundations and airfields that would surpass any on earth. They were building an airfield – by God! – and a big one, too. The biggest in the world, they were told.

There was a pride in their work, along with a sense of importance that spurred the workers on with their creation. Rumours still travelled of a war stirring "over there."

Almost everything was a first for these working men from the outports. All of the men had come from simple, isolated towns that clung to the rocky coves and fjords stretching along the many secluded bays. Few if any of their simple homes were equipped with central heating. Their mostly saltbox-style houses possessed one lone stove in the kitchen that took care of all their cooking needs, but was usually doused with the last of the teapot contents before the tired housewife sought her cold upstairs night bed, plunging the home into a breath-stealing cold.

Here in the new "townsite" of Gander, the men slept in heated tarpaper bunkhouses and ate in huge, always warm mess halls. They were served three hot meals a day, and sometimes more – cut and come again. Wondrous food. Great roasts of beef, pork, chicken, and sometimes turkey. Rashers of bacon and countless morning eggs. Endless desserts and sweets for the taking. These simple, hard-working people were amazed, and showed their gratitude with pick and shovel and saw and ringing hammer as they created the new airport town of Gander out of the woods and into twentieth-century history.

From the conception of the North Atlantic Civil Agreement by the countries of Newfoundland, Canada, England, and Ireland in 1935, and the actual start of construction with just forty men in 1936, came great strides in international co-operation. The location of Hattie's Camp was the closest point to the great circle air route crossing the North Atlantic to England. Just two years later, on January 11, 1938, Captain Douglas Fraser landed his Fox Moth onto the gravel airstrip. None of the pioneers of this operation could know that from such simple beginnings would come many historical and world-changing events. The airstrip at Gander would also change, forever, countless lives.

* * * * *

MY FATHER WOULD LIGHT smudge pots – kerosene-burning torches – when a plane came after dark: a dull, flickering yellow flame that burned and smoked and stank and guided the rumbling planes to the safety of earth. With a five-gallon can of white paint hanging from one hand and a six-inch paintbrush in the other, he painted, without the aid of stencils, the first numbers on the concrete airstrip. From the steady stream of cargo planes he would help unload bombs and store them in concrete underground bunkers.

One thing the workers didn't like very much was the "privy" system. Outside, at the back end of the bunkhouse, which housed as many as 100 men or more, was a slanted-roofed building. Inside was a long, raised wooden bench over a deep, stinking pit. Here the men waited their turn to sit side by side with as many as fifteen others and do their daily business with no privacy, magazines, or toilet tissue. It was not the most hygienic setup.

Over the years this airport town of Gander would have its image stamped on every cockpit chart across North America and beyond. The Ferry Command created by the Allies used Gander as the jumpoff for fighter planes to fly across the great circle route in defence of Mother England. This service, which would prove to be an immense success, would not only establish Gander as a world-class airport locale, but would also do more to ensure victory in the Second World War than any other effort during that terrible half-decade of the 1940s.

Slow-moving, zigzagging freighter convoys with fighter planes lashed to their decks came under relentless fire by German U-boats using the "wolf pack" attack formation. When ships crossing the Atlantic were torpedoed, and even just one sky-fighting machine was sent to the bottom of the sea, the loss was keenly felt over British skies. However, Gander gave the Allies near total air superiority over the Nazis. Eventually the decision was made for the planes to cross the ocean on their own power. The constant supply of new machinery flown "across the pond" from the airport town to England far outstripped the destruction of their aircraft over war-torn Europe.

The island of Newfoundland, on the edge of the continental shelf, was of course not only the springboard for east-flying North Atlantic aircraft, but also the welcome destination for all westbound planes coming in over the ocean. It was the crossroads of the entire aviation world, the aerial gateway to the North American landmass, the gas pump of the Atlantic.

The years progressed. From its humble wilderness beginnings, the town of Gander grew and its airport evolved. It never got any bigger. It was already big enough to be designated as an alternative landing spot for NASA's space shuttle. The aircraft that crossed the same ocean got much bigger, and faster, too. Their size and speed meant they needed more fuel, but the same modern technology made them more fuel-efficient. The huge airplanes, now so big they took on the names of ships at sea – liners – didn't have a reason to stop on the edge of the Atlantic anymore. The world had become much smaller.

CHAPTER 1

WEDGED LIKE AN ANCIENT, well-worn arrowhead between the Gulf of Suez and the Gulf of Aqaba, the Sinai Peninsula holds back the blue waters of the Middle Sea from mixing with the narrow Red Sea, which like a great swimming serpent flicks its forked tongue east and west around the southern coasts. In from the warm Indian Ocean to the south and detouring from the cold Atlantic of the north, great oceangoing ships come to the Sinai, its canal at Suez conjoining the earth's vast north and south salt seas.

Away to the west, inside the seven deltas of the mighty Nile River, and south, past the arid deserts of northern Egypt, the dark continent of Africa funnels its great mysteries and exotic smells of timeless forest and endless fruited plains north, to the troubled lands of Abraham and all of his get. Cush, son of Ham, grandson of Moses, inherited the land and gave it his own name – the land of Cush. It had been fought over and conquered, plundered, and envied for generations, until Moses, with nothing more than a wooden staff and Divine assistance, led his people away in search of a wondrous Promised Land that he would never see. From Succoth and across the God-given parting of the northern Red Sea, and on into the Sinai's desert of Shur, Moses led his people in that "great and terrible wilderness" for forty years.

Driven into the tumultuous centre, between the Earth's largest continent of Asia and its second-largest, Africa, the lone stepping stone between Africa and Asia is Sinai. The only way into Asia or Africa by land is across this peninsula. The ancient Persian empire, which stretched from the Arabian Sea in the south to the Black Sea in the north, fought and finally conquered the northern sea edge of the Sinai as it plundered its way west into Africa and feasted on the incredible riches of the Nile delta.

Eight hundred kilometres south from its northern mouth, the Nile makes a great bend toward the hot, arid east, as if wanting to merge with the Red Sea. Here in the centre of this oxbow flow, before the first cataract of the great river, kings had lived in their private valley in the glorious city of Thebes, on the east bank of the river many of them worshipped. At the end of its northern flow, the Nile empties its seven mouths into a vast, flat delta. The salt of the middle sea slows and cleanses the fresh water, which travels up from the mysteries of Africa. The river deposits untold tons of soil stolen from the endless valleys from whence it came, all along the ageless flood plains and blessing them with its yearly rent of tumbled fertility.

Wise Saladin, too, the Kurdish Muslim sultan of Egypt and Syria, nemesis of the English crusader King Richard the Lion-Hearted, came with his army, riding their magnificent "drinkers of wind," loving this land with its dark-eyed women. But through years fraught with countless battles, it was the Sinai that held invaders together or kept them apart; it was the referee of the Middle East, the ace up Egypt's sleeve. This land of legends was the birthplace of the world's three major religions – Judaism, Christianity, and Islam – all brandishing blood-stained swords in the name of their respective gods.

The most aggressive warmonger of them all, the Great Alexander of Macedon, young son of Phillip II, marched at the head of his terrible army across Sinai, bearing his Greek empire westward. Little Napoleon Bonaparte of France with huge ambition fought his way across this land, the dry desert absorbing the blood of his fight as easily as it did that of ancient warriors.

Modern wars were waged here, too. In July of 1942, the battle of El-Alamein, between the Axis forces of Germany and Italy under the command of Irwin Rommel, and the armies of Britain, India, Australia, South Africa, and New Zealand, raged across the western desert of Egypt. Before it was over, more than 13,000 lives had ended, with a few covering shovelfuls of hot desert sand to mark their passing.

Besides its strategic geographic position, historically Sinai was fought over for another, more colourful reason: a treasure valued more than all others among the ancient warriors; a small blue gem – turquoise. It was hoarded by kings, used as currency, bartered for, battled for.

Sinai, where no river flows and no blue lake of fresh water shimmers in the heat of the terrible midday, held the precious jewel of vanity beneath its arid bosom. Turquoise, a hydrous phosphate of copper and aluminum, is blue and sometimes green in its natural colour, its name suggesting the land of the Turks. The valued gem wasn't found anywhere in Turkey, but rather was brought to its marketplaces by Venetian traders who had haggled for the precious metal in Sinai. From the hills of sandstone, limestone, and basalt came the most translucent turquoise to be found anywhere. It was laboriously hammered from its inorganic bed by sweating, muscled men and hauled out of the world's very first mines.

Sinai's mark on antiquity is only strengthened by its modern history. On November of 1869, a canal was finished along its western edge. The opening of the Suez Canal merged the salt waters of the great northern and southern seas. The canal contained no locks, but allowed the blue water of the Mediterranean to flow at sea level, freely into the Red Sea. Heavy-laden, fully-rigged clipper ships with exotic spices, silks, and scented perfumes from the far Orient suddenly didn't have to venture with billowed sail around Africa's most southerly Cape Agulhas. Nor did they have to round the Cape of Good Hope, which all sailors dread, farther along that windswept coast.

The manmade trench through the western flank of the Sinai had shortened not only north- and southbound traffic, but almost all east- and westbound vessels in that part of the world forever. Flooding Sinai's soil with new sea water only increased its strategic importance worldwide. Long used to being under the envious eyes of the ancient conquerors, it had now come under the constant gaze of the modern ones. Its narrow waterway, making previously long ocean treks much shorter, came under an international law which states in part: "It can be used in time of war as in time of peace, by every vessel of commerce or of war, without distinction of flag."

At the American president's luxurious retreat called Camp David in the state of Maryland in the United States, on the warm spring morning of March 26, 1979, thousands of miles away from the deserts of Egypt, the fate of the Sinai Peninsula's ribbon of salt water was again sealed. Prime Minister Menachem Begin of Israel and President Anwar Sadat of Egypt sat down with the President of the United States, Jimmy Carter, who sponsored the event, and the modern-day Camp David Accords concerning the Sinai were signed.

The Multinational Force and Observers (MFO) were created in 1981. It was a coalition of world armed forces to ensure peace between Israel and Egypt. The commercial viability of the Suez Canal had to be protected, regardless of Middle East differences. Several countries were involved, the United States of America being one of them. The 101st Airborne battalion out of Fort Campbell, Kentucky, was stationed in the Sinai to ensure that its country's interests were protected.

The 101st Airborne Division of the American military had been activated on a warm August 15 of 1942. This select band of brothers immediately took as its motto "Rendezvous with destiny." Their nation was already involved in a terrible worldwide conflict. The warring fields of Europe would soon provide their rendezvous, as well as reveal their destiny.

They took as their fighting mascot their national emblem, the bald eagle. It had already been used during the American Civil War by an infantry regiment from the state of Wisconsin, whose favourite name for the emblem was Old Abe. Bearing the magnificent, white-headed raptor with open beak proudly upon their shoulders, the 101st "Screaming Eagles" – air assault – went to war.

* * * * *

BY JUNE 1944, ENGLAND had been battered by nearly five years of war. Until now it had been a largely defensive war for the Allies. Only twenty-one miles across the Dover Strait on the occupied coast of France, the dogs of Hitler bayed. The Germans had bombed military sites and war machinery factories day and night. They switched their tactics, believing a direct attack on the British people would defeat them. Night after night, the city of London bore the terrible wounds of falling, screaming bombs. The senseless slaughter of hundreds of innocent men, women, and especially children became the rallying cry of a nation that would never surrender. The mistake the Germans had made in changing their bombing runs would cost them dearly. It was time to take the war to Hitler.

Four years previous, on June 6, 1940, the last of the 333,226 Allied troops, after a failed British invasion, waited for rescue on the beaches of Dunkirk in northern France. It was the greatest retreat in British history. Before the German forces could reach them, every boat afloat on the south coast of Britain answered the call. Back across the salt channel, where Julius Caesar had led his invading horde thousands of years before, the retreating army sailed home to lick its wounds. Every last one of the Allied forces was ferried to safety across the Dover Strait to England.

Under the cloak of night on June 5, 1944, they returned. The greatest amphibious attack force ever organized gathered just outside the surf line of Normandy's beaches, on the north coast of France. They were waiting for the dawn of June 6. Their revenge would be bittersweet and complete. The Allied forces were about to take their fight into Germany's unprepared lap – Berlin. This time there would be no retreat.

The 101st Division of the American forces was a very colourful group. Each member wore the clearly defined mascot, Old Abe, upon his shoulder sleeve. They identified their separate divisions by card suits, bearing the Clubs, Hearts, Spades, and Diamonds symbols on their helmets. They were always at the forefront of the battle. This battle would be no exception.

In the black hours between 1:00 and 2:00 a.m., three divisions of paratroopers dropped behind the German lines of occupied France. The Screaming Eagles were one of them. When the paratroopers of the 101st jumped from the English C-47s into that clouded night, they fell down into a battleground of Biblical proportions.

* * * * *

LIEUTENANT COLONEL ROBERT COLE yelled, "Time to fly, Eagles, hit the silk!" and leaped out of the belly of the huge plane at the head of his "stick" – the British word for a group of parachute jumpers. His voice was swallowed by the backwash roar from the props. He fell free a couple of hundred feet and pulled the ripcord, stopping in midair and snapping backwards by the sudden pull of the inflated chute billowing above his head. He looked at his watch. It was 0053 hours. They were right on schedule. Glancing up, he watched the sky become dotted as if by small white puffs of cloud as his men exited the plane. The sky was quiet when the aircraft droned away. The strings holding Robert to his chute whined and hissed softly as he glided toward the ground he couldn't yet see. There would be no dropping "like a thunderbolt from the skies onto their enemies" on this night, he thought.

Stealth was an asset now in the black night that secreted them. The navigator of the plane was supposed to have delivered them above and just inside German lines, at the base of the Cotentin Peninsula, near the city of Cherbourg. Cole hoped the long-gone navigator had been at the head of his class.

As he fell, smells from the activity on the coast below and farther to the north drifted up to him. Diesel fumes from thousands of gathering ships rose on the night currents. Operation Overlord, years in the planning, had begun. Before this day was done, 9,000 ships, carrying more than 155,000 men, along with thousands of pieces of different vehicles, would come ashore on the unsuspecting beaches of Normandy. Before Operation Overlord was over, a million men and more than 177,000 vehicles, all ready for battle, would be landed along this coastline. As the coming dawn ended the night, nearly 5,000 warships would begin bombing the land ahead of the advancing forces. The gathering invaders sailing from the ports of England would put the old Admiral Nelson's floating armada to shame.

Cole pulled on the risers and suspenders of his chute. They gave him a small measure of direction as he fell earthward. All around him men hung in the air, their arms raised over their heads, hanging on to the lanyards connecting them to their fragile silk umbrellas. No one spoke. Not even a cough escaped the night warriors as they dropped from the sky. Except for the swish of air rushing through the many risers like the wings of a flock of night birds, all was silent.

Then the gunfire started, at first no more than faint popping sounds from below. A dense fog came up at the paratroopers as they fell. Then, through the mist, quick bursts of flame, followed by the deadly sound of gunshots, filled the rising landscape all around them. Robert heard a scream. It came from above his straining chute. His men were all around him, falling silently.

The greatest fear for a falling paratrooper was to be spotted and shot at from below. Robert shivered and pulled his legs up as far as he could, trying to make his body smaller. A sky-lining black figure topped by a billowing white canopy, coming ever closer to a rifleman waiting on the ground, was easy pickings for the enemy. He knew of many who had died in such a manner. Their bodies riddled with bullets, their heads slumped on their chests, they had dropped and crumpled, dead before they hit the ground. There were stories even more horrible. Soldiers had seen the parachutes above dropping paratroopers shot to ribbons, the men falling helpless, still alive, like flying geese whose wings had suddenly folded. Other accounts told of paratroopers who had been intentionally shot through the legs and even gut-shot as they closed with the ground. They were dragged screaming away from their drop zone, to undergo unheard-of agonies of interrogation.

Robert dropped to a stumbling, staggering run onto French soil and watched as his battalion descended out of the night one by one behind him. They dropped like black, shapeless forms. All around them now was a staccato of gunfire, the night stabbed with rapid yellow flashes from hundreds of rifles.

Robert G. Cole was a world away from home. He was born in Fort Sam, Houston, in the state of Texas, U.S.A. He knew that he and his men were taking part in the greatest assault in history. He was proud to be a part of it. Their orders from the commanding officer of the 82nd Airborne Division seemed daunting. They were to destroy the Germans' coastal artillery at Saint-Martin-de-Varreville. Capture strategic buildings in Mezieres. Take control of the Douve River lock at Le Barquette. Capture the two footbridges crossing over the Douve at Le Port. Destroy the vehicle bridges at Saint Côme du Mont. Make the entire Douve River valley safe. Destroy German communications. Set up roadblocks. Establish a defence line. Make drop zone areas safe. Capture the city of Carentan. Connect with others of the 82nd Airborne Division. When the commander had finished giving the orders, he asked if there were any questions, his voice a growl. There were no questions asked.

There were many airborne divisions dropping onto French soil this night. The 101st led the way. When dark ended that same day, unbelievably, 23,000 young men of the Allied airborne divisions had become casualties. Robert Cole was not one of them.

Five days later the 101st were fighting for control of the last bridge crossing the Douve River on the road to Carentan. Cole's company had been pinned down under a murderous fire for more than an hour. Many of his men had been killed. Under artillery, mortar, and rifle fire, they found themselves in a perilous situation. Cole's company, battle-weary and fatigued, and suffering from lack of proper food, water, and sleep, were about to be defeated.

Cole peered up over the edge of the ditch and along the road. The air was so thick with cordite it burned his nose. He coughed and spat dust-coloured phlegm from the back of his thirsty throat. He looked back at his men. A steel helmet on the ground caught his attention. He was about to yell to the soldier sitting near it to put it back on, when he saw a thin line of blood trickle down the man's face. The soldier was sitting on the ground facing Cole. The young man's eyes were open, but they would never see again. The blood ran slowly out of a small, neat hole above the soldier's left eye, leaving a black trail the length of his dusty, bristled jaw.

Several more soldiers were moaning from gaping wounds. One man's lower abdomen was twitching pitifully, the upper body still. Two soldiers lay sprawled where they had fallen. For them the battle was over.

Cole's men peered up over banks and through bushes. They were all staring at him, waiting for his orders. It was decision time again. Cole yelled to his men to fix bayonets. "It is time for our rendezvous with destiny!" he shouted like a madman.

And standing to his feet, the man from the fragrant, lush plains of far-off Texas led his men in a zigzagging run across a stinking, war-torn field in France. Behind him came a blood-curdling battle cry as every one of the Screaming Eagles dropped like a thunderbolt upon their enemies. The Germans were shocked to see what was coming at them. They stared for a moment in stunned surprise. They thought they had beaten down the Americans. It was a big mistake.

For his decision and act of valour against all odds, Lieutenant Colonel Robert G. Cole received the Congressional Medal of Honor – posthumously.

CHAPTER 2

WALKING DOWN THE STEEP ramp from the Arrow Air plane, Darrin Patrick Brady was surprised by the full force of the hot desert air that pressed against his face. Though he'd been forewarned by his officers and others who had been here before, he was totally unprepared for the sudden blast of stifling heat. The air was breathless, not a stir of wind to fan even the faintest whiff of cool air across his lean face. The sweet, sticky smell of spent jet fuel hung in the air.

The land shimmered in a high-reaching haze as far as he could see that seemed to hold it prisoner in a grey, bowl-like depression. It was Darrin's first time here, his first glimpse of the largest city in North Africa – Cairo. He was part of the 101st Airborne Division deployed with the MFO, the Multinational Force and Observers created in 1981 to ensure peace between Egypt and Israel.

It was June 1985, Darrin's first day of a six-month tour in the Sinai. In fact, it was his first military tour anywhere. He was nineteen, one of the youngest soldiers in his division. His own rendezvous with destiny was only beginning.

Darrin was from Brunswick, Ohio. With a population of close

to 30,000, it was the largest city in the county of Medina. His home state had a northern border along the coast of Lake Erie, named after the Erie people who once lived along its southern shore and called the inland sea their own. Ohio was the Native Seneca word for "large creek." Because of the many buckeye trees growing in the lush state, it was known as the Buckeye State. Darrin's soldier friends of course nicknamed him "Buckeye."

Coming in over the blue Mediterranean, the sleek Arrow Air plane had banked south away from the salt sea, flying low over the dirty-looking Nile River before touching down on Egyptian soil at Cairo. Darrin wondered about the name. There was a town in Illinois called Cairo. The Ohio River, which gave his home state its name, poured its huge volume of silted fresh water south toward the Gulf of Mexico. Where the Ohio gave up its 981-mile-long flow along with its name, to that mightiest of America's rivers, the Mississippi, the city of Cairo, Illinois, flourished at the confluence of both south-flowing rivers. The waterways that had borne families of Iroquois, Osage, Omaha, Ponca, Kaw, and Erie people over them in their magnificent wooden canoes, now carried foul-smelling steel barges to and fro.

This other Cairo, of which Darrin had suddenly become a part, boasted the longest river on earth. The fresh water of the Nile flowed north for 4,150 miles before blending with the salt of the Middle Sea. The river and the city name was the only connection Darrin felt to his distant homeland. This desert land was dull and drab, with only a few struggling patches of green that he could see. The city of Cairo, on the east bank of the Nile, glared up out of the desert like a Bedouin tent in a sandstorm, its trendy modern towers standing in contrast to its humble beginnings.

Darrin ran a hand up through his thick, dark hair. It was cut to regulation length. Raising his arms above his head, he yawned and stretched his six-foot frame. It had been a long day. The time changes would take some getting used to. The long flight from the States had made it seem like he was always flying into another country's morning. He needed a shower, a cool drink of water, some sleep, and something to eat that tasted better than airline food. Even Army fare would be better. Darrin and the rest of his battalion would have no chance for much personal attention yet. They still had a ways to go. They were headed for the south camp of the MFO, situated in the southern end of the Sinai Peninsula.

Darrin had read up some on Egypt's history. He knew Cairo was referred to as the city of a thousand minarets. He listened for the call to prayer that was supposed to come from the sky-lining towers no less than five times a day. Sprawled like a great oasis across desert sands and rising above ancient, dun-coloured structures, Cairo rose in all of its modern splendour. With over six million inhabitants in this largest city in North Africa, the jewel of Lower Egypt, many of them Muslims who followed the Islamic religion, Darrin figured at least one muezzin would be crooning his message to the faithful. He didn't hear anything.

Maybe they aren't any better churchgoers than me, he thought. Darrin wasn't part of any religion.

Music started somewhere, the sound sharp and clear. It was not coming from a minaret. A voice sang over and over – *Born in the USA*, Bruce Springsteen's song, #9 on the American charts – coming from a ghetto blaster carried by one of the soldiers. The sound from home was motivating, uplifting to the American soldiers. Someone cheered, the repeated line from the song bursting from his throat. Several more joined in – "Booorn in the USA, booorn in the USA!" – their voices rough and swelling with enthusiasm, with pride. They were very patriotic. Darrin joined in with the singers, his strong, deep voice blending in with the rest. Their battalion officer looked their way, smiled, and looked away. He was pleased with the morale of his men. *Esprit de corps* was important to the 101st.

Darrin liked Bruce Springsteen's music, all right, but his favourite music band was Queen, especially their 1980 hit, *Another One Bites the Dust.*

Lots of dust to bite around here, Darrin thought.

Springsteen's song ended and another one came on, *Eye of the Tiger*, by Survivor, a hit song from 1982. It was another one of Darrin's favourites. His sister liked the same kind of music, too. Darrin smiled, thinking of Shelley.

* * * * *

SHELLEY WAS SIXTEEN WITH a full head of blonde hair, and blue eyes that always smiled for her brother. She was his only sister and he was her only brother. They were good friends. Shelley respected her brother's decision to enter the Armed Forces, even though she wasn't sure if he should make it a career move yet. Darrin wasn't sure, either.

There was one thought that bothered Shelley, though. Darrin had sent home a few pictures from the Sinai. One of them showed her brother naked from the waist up, his tanned, well-muscled body standing firm, defiant. Draped over his broad shoulders was a military harness accoutred with army gear, pouches filled with soldier stuff, a pair of binoculars, and what looked to Shelley like a holstered pistol. In the centre of his chest his dog tags glistened like a shining target. Shelley cringed at the thought. Darrin was wearing a camouflaged baseball-style hat, his eyes covered with a pair of dark sunglasses, hiding his brown eyes. Shelley thought her brother looked like Rambo.

Another photo taken at night: her brother in the glare of an armoured vehicle. Dressed in full desert camouflage gear, with red company beret draped over his right ear, shirt sleeves rolled up, Darrin looked ready for anything. Clutched firmly in both his hands was another gun. It was an evil-looking thing, long, dark, and deadly. Shelley thought the rifle looked out of place in her brother's hands. She had never seen him with a gun. Darrin wasn't a hunter; he didn't even go fishing. She didn't believe her brother could kill anything. They were on a peacekeeping mission "over there," she reasoned. It wasn't like Darrin had gone to war. She often wondered what Darrin would do if he actually had to shoot at someone. Shelley hoped her soft-hearted brother would never be put to the test. Staring at the photo, she noticed Darrin wasn't smiling.

* * * * *

DURING THE NEXT FEW months, Darrin would think of his fair-skinned sister often. He wondered if there were any Sinai women with blonde hair and blue eyes. He had not seen one. There was another girl in Darrin's thoughts, too, one who would dominate his mind and cause him to squirm in his night bunk before finally finding sleep. Sue Sinclair was Darrin's high school sweetheart. She was now his fiancée. Sue was as blonde as Darrin's sister, and very beautiful. Darrin and Sue had clung to each other when they parted, pledging their love over and over again, the young lovers not realizing how long six months apart could really be. By the time September rolled around, Darrin and Sue were having problems. Their love hadn't been as strong as they had imagined.

The days and months wore on for Darrin. The heat never let up. He was told he would get used to it, but he never did. Even the wind was hot. It seemed to funnel down from the barren hills and sweep across their camp in heavy gusts that never cooled the skin. Sometimes the winds whipped up sandstorms that were worse than a winter snowstorm back in Ohio. Darrin filled his off-duty hours with exercise. Despite the heat, he intended to keep in good shape. He had worked out regularly at home, and besides the daily required physical activities, the Army provided everything needed for a good cardio workout here in the Sinai. Darrin was in the best shape of his life. He was eager to try new things, glad of his experience in a part of the world he never thought he would see. He liked the way he felt, always energized, full of life. He was actually enjoying his Army life, and was proud to be a part of the 101st battalion. Maybe he would make the Army his life.

Darrin was under no illusions that the forces he was involved in would always be in a peacekeeping situation. He was well aware of world events that could send the 101st to heavy armed conflict. It would only require an order given from their commander, and they would put all of their stringent training to deadly use. They were always kept on a knife's edge of preparedness, always battle-ready. Their guard was never down. It was their history, their pride: to a man, they were honed for their individual and collective rendezvous with destiny.

* * * * *

ONLY ONE YEAR AGO, in June of 1984, in South America, the Salvadorian Air Force had dropped 3,000 tons of bombs on civilian people. The raids had killed more than 2,000 innocents. The bombing was classed as the fiercest air war in the Americas. The bombs were made in the United States. Although the incident was relatively close to home, the media had ignored the story.

Darrin was aware of another unbelievable atrocity that had been inflicted upon faultless people. This one was on a different continent. In June of this year – his battalion were aware of it when they were deployed – Amnesty International reported to the world a horrific human rights violation. As many as 200,000 East Timor people, all of them civilians, had died as a direct result of Indonesian aggression. The killings took the lives of one third of the population of East Timor. The American press refused to publish that apocalypse.

In that same month of 1985, another act of terrorism was boldly played out for all the world to see. Though the consequences were not nearly as severe, it showed everyone the type of war that would be fought by some rising world powers. It would be a war against the very ones whom elected governments and fighting soldiers alike were sworn to protect – the innocent. A deadly, unpredictable war unlike anything our planet of countless wars had ever seen. It was only the beginning of the wars of terror.

WHERE EAGLES LIE FALLEN

On June 14, TWA Flight 847 left Cairo, Egypt. It was a Friday morning. The plane was scheduled to stop in Athens, Greece, then proceed on to Rome, and from there to London, England. The plane was piloted by fifty-eight-year-old John Testrake, an American. On board were 153 people, consisting of passengers and crew. They would all be forced to endure a three-day life-changing event.

Hijackers took over the plane shortly after takeoff from Cairo. They ordered the pilot to change course to Beirut, Lebanon. The plane was given a reluctant clearance for landing in Beirut, where the terrorists allowed a few women and children to leave the aircraft. The usurpers ordered the plane away again, this time west, all the way to Algiers, North Africa. It took a personal call from President Ronald Reagan of the United States to Algiers president Chadli Bendjedid before the plane was allowed to land there. More passengers were allowed off. Negotiations failed. The plane took off again.

The TWA aircraft flew back to Beirut, landed a second time, and the passengers were further traumatized inside the sweltering airliner. The hijackers ordered the release of Shiite prisoners held in Israel, Cyprus, and Kuwait jails. They also demanded an international condemnation of the United States and Israeli governments. Their demands were not met.

To show they were sincere in their endeavours, the radicals walked the length of the plane. They pulled a man, an American Navy diver, out of his seat and beat him severely. They placed a gun against his right temple and pulled the trigger, then dumped his body onto the tarmac. The passengers aboard the plane knew real fear then.

The extremists then instructed the pilot to fly his plane back to Algiers, where they let off fifty more of the passengers unharmed. The airliner needed more fuel, but the Algiers airport refused the fuel without payment. The skyjackers grew infuriated. Uli Derickson, a cabin attendant from Germany, the only non-American member of the crew, paid $5,500, using her own Shell Oil credit card, for the jet fuel. Once fuelled, the plane returned to Beirut after negotiations again failed. This time Beirut refused permission to land. The pilot pleaded with airport authorities, informing them he had only five minutes of fuel left. He was finally allowed to land in Beirut again.

Back in the United States, the government dispatched the elite anti-terrorist Delta Force. This was a crack, no-nonsense team ready to deal with anything. The game these evil men had started was about to spread onto a world field. After several more threats of terror, they finally released the remaining hostages from the plane, blew up the aircraft on the apron of the Beirut airstrip, bolted into a nearby Shiite neighbourhood, and disappeared.

October 1985 was not a good time to be travelling in the Mediterranean area. On the seventh of that autumn month, four heavily armed men, claiming to represent the Palestine Liberation Front, took control of the huge Italian cruise liner *Achille Lauro* off Egypt's coast. The ship was sailing from Alexandria, Egypt, east to Port Said in that same country. Terrorism had taken to sea.

They ordered the ship to dock in Tartus, Syria, and demanded the release of Palestinian prisoners in Israeli prisons. They were refused. Again the terrorists wanted to send a clear message, and again their proclamation was a deadly one. Again, an American innocent was their scapegoat. Sixty-nine-year-old Leon Klinghoffer was a travelling American citizen. Leon was disabled and had to use a wheelchair to get around. He was ruthlessly killed. His body, limp in his wheelchair, was thrown over the side of the *Achille Lauro* into the warm waters of the Middle Sea.

In Port Said, the hijackers relinquished the ship after being assured safe passage to Tunisia aboard an Egyptian airliner. President Reagan was not pleased: another American citizen had been killed by international terrorists. He sent direct orders to the American aircraft carrier *Saratoga*, which was already in the area. The great ship came into the wind. F-14 Tomcats raced across her lifting runway, lurched a bit when they ran out of deck, and then roared away from their floating hangar, their vapour wakes shimmering. The Egyptian airliner was rapidly overtaken, and its pilot ordered to fly to the NATO Naval Air Station Sigonella base in Sicily. The criminal masterminds were arrested by the Italians.

On November 23, the event still on everyone's mind, terrorists hit again. An Egypt Air Boeing 737 was seized and forced to land on the island of Malta. The Egyptian armed forces took back their plane, but fifty-nine people died as part of the bargain.

* * * * *

DARRIN TRIED TO FORCE the thoughts of terrorists and their cowardly method of fighting from his mind. Though he never talked about it much, he was convinced that future wars would be fought this way. He couldn't see how such a war could ever be won against those who used such tactics.

Darrin spent some of his leisure time drawing. He had always been good with his hands, at almost everything he tried, but he surprised himself when he learned he could actually draw. Family members who saw his work agreed he was good. His sketches mostly consisted of swashbuckler-pirate faces and resplendent Indian heads he saw in magazines. The ads were always the same: draw the face, submit it to the address provided, and you could be part of a correspondence art school. Darrin drew the faces, but he never did submit any of his work. He was especially good at drawing animals and birds. He had sketched a turtle once; it was his sister's favourite. His handwriting was an art in itself, the lines free-flowing, strong, and distinct, his artistic bent coming through with ease. Shelley always envied his ability and told him his handwriting skills were too fine for a man.

Darrin was on a motor patrol with one of his company groups one sweltering day in late September. The surrounding area was sparse and the air was hot, thick with the heat rising from the parched land. There were some green shrubs growing among scattered dust-covered rocks, and a few stunted trees, their shrivelled leaves looking as if they would escape the branches any minute. The company had patrolled the same road many times.

A tree taller than the others caught Darrin's eye. Something was hanging from several of its naked limbs. The truck slowed and then stopped at his request. Darrin wanted to investigate the strange-looking leaves. The commanding officer of the unit gave him a firm no; there are always landmines. Instead, Darrin pulled the powerful binoculars out of their holder.

He believed the tree that loomed into his sight was an acacia. Its claw-like branches were armoured with dozens of long, sharp thorns. The acacia tree was capable of growing in sandy terrain such as the Sinai's. It is believed to be the plant of "burning bush" fame, which the Hebrew Moses saw and wrote about in the Christian Bible. The same Bible referred to the burning of the acacia wood as giving off the finest incense. Hiram Abiff, master builder of King Solomon's tomb in Jerusalem, used wood of the acacia. Some scholars believed a thorny branch from the acacia adorned the head of Christ on the day of His crucifixion. The gnarly tree also signifies resurrection and immortality.

Impaled upon many of the acacia's razor-sharp thorns were the carcasses of small songbirds. Some of the bodies stood upright while others were skewered upside down, their limp tail feathers shifting in the breeze. A small lizard hung from the tree, its short legs splayed open, its gaping mouth looking grotesque through the powerful lens. Darrin didn't know what he was looking at. One of his team was an amateur outdoorsman. Michael knew just about everything concerning wildlife back home in the States. He was spending his free time here learning about the different species he was encountering.

It was a shrike's pantry, Michael told him. The shrike, a hawk-like bird, dined on small birds, lizards, and even big bugs. The bird would keep hunting, even when its stomach was full, and hang its prey, usually still alive, upon the thorns of the acacia, to be eaten whenever the bird was hungry. Darrin was fascinated by the shrike and its method of storing food. He suddenly felt the urge to sketch the bird.

Weeks passed, but Darrin got only fleeting glimpses of what he believed to be a shrike. The end of November was near, and now it was barely two weeks before going home. One evening, while the sun slid down behind the quiet hills of the Sinai, the artist in him stirred at the sight of the evening colours on the horizon. Red and pink and deep purple hues stained the sky and blended together like an artist's pallet.

Darrin thought of the shaky relationship between him and Sue. Maybe spending some time in their old favourite places would bring them together again. Neither Sue nor Darrin were sure anymore. He didn't know if this time spent apart would spell the end of their relationship.

A bird appeared on one of the metal fence posts surrounding the compound. Darrin hurried for paper and a pencil. When he returned, the bird was still there, as if posing. He hurried to trace the hawk while the light lasted. The handsome bird looked like a falcon. It sported a black raccoon-like mask across it eyes that stretched around the back of its head. When he finished the drawing in the dim glow of the desert twilight, the hawk was still on the post, as silent and still as before. With the light gone, the bird stood in a grey-black silhouette against the sky, which was just fine with Darrin – he didn't use colours in his drawings. Soon after the light left the sky, there was a flutter of wings and a shadow jumped into the night sky. Darrin's splendid model had gone.

Curious to know whether or not the bird had been a shrike, Darrin showed his drawing to Michael, his knowledgeable friend. Michael looked down at the drawing, then back up at Darrin. He asked if Darrin had sketched the bird himself. He said the drawing was very good; Darrin had captured this particular bird's silhouette perfectly. It was the most telling feature when trying to identify the shrike, Michael told him. Darrin was pleased, but tried not to show it.

However, Michael told him that it was not a shrike he had

drawn after all. The bird's markings he had so skilfully drawn were unmistakable. It was the magnificent raptor, Michael said, which spent its winters on the Arabian Peninsula. Darrin's first impression had been right. The bird was a saker, a desert falcon. For ages, desert sheiks in billowing silken robes had released these magnificent raptors from their wrist-worn leather jesses and marvelled as they soared and hunted, returning to outstretched arms with their bleeding prey.

Michael had been hoping to see at least one saker before leaving. He asked Darrin if he would be willing to part with the sketch. Darrin was flattered. He told Michael he wanted to touch it up a bit more, and maybe he would give it to him then.

* * * * *

AT LAST THE LONGED-FOR day came. It was time to leave. Darrin's first six-month tour of duty in the Sinai was behind him. He was very pleased with the experience. He had learned much about the Middle East and its ways, about his regiment, about living and working with his fellow soldiers. Darrin had learned about himself. He was not the same man who had come here six months ago. He was ready to go home.

Darrin had most of his gear packed before departure day. He was excited, the jubilation of going home after so long an absence pumping through his system. Taking one last look around, he was surprised to discover he had not packed his drawing of the saker. His duffle bags had already passed through customs and were on their way via truck to Cairo Airport. He neatly folded the sketch and placed it in his shirt pocket.

CHAPTER 3

Darrin "Buckeye" Brady's regiment left the south camp of the MFO on December 10, 1985, for a much-deserved trip home. His group flew from Sharm el-Sheikh Airport, near the town of Ras Nasrani, to Cairo in two Egypt Air planes. Their luggage went by truck, to meet them at their destination.

The second flight of Egypt Air, with the last of the 101st aboard, arrived at Cairo International Airport on the evening of December 11 at 2:00 p.m. Their replacement crew from the States arrived at 5:35 p.m. aboard an Arrow Air craft. This fresh crew of the 101st Airborne Division, many of them looking as young as Darrin, were scheduled to go through the same procedures as he had begun six long months ago. Many of them had the pale winter skin of the northern states. Darrin smiled to himself. The Sinai heat would soon change that.

An electrical problem aboard the plane delayed their flight by an extra thirty minutes. The plane left for Cologne, Germany, at 10:35 p.m. Flying into a night sky, the 101st headed for home.

Arrow Air Flight 1285 was registered and identified with the number N950JW. The DC-8-63CF jetliner was not a military plane. It was a long, white, sleek-looking aircraft built by the McDonnell Douglas group of companies. The Arrow Air company was founded and owned by a Native American, George Batchelor.

George was born in Shawnee, Oklahoma, one of the Great Plains states of America, in 1905. George had always been fascinated with flight and planes. Not only did he want to fly, but own his own airplane. He would become very successful at both. Even the death of his wife and young son in a horrific plane crash would not quell his love of flying. He moved to Compton, California, in 1947 and there established Arrow Air.

In 1964, George Batchelor moved to South Florida. There, he was considered an aviation pioneer, as well as a forerunner in the Latin American air industry. George had a soft heart for children. He gave considerable amounts of his own money to organizations for homeless children.

Before this date in 1985, Arrow Air boasted a record of having flown more than one million passengers to 245 destinations in seventy-two different countries. The airline had been approved by the United States military for troop transport the year before. In 1985 the company had secured a $13 million contract with the American Department of Defence. George Batchelor was very proud of the fact that Arrow Air had never had a fatal accident.

High above the eastern end of the Mediterranean Sea, north over the narrow island of Crete and past fabled Greece itself, birthplace of Socrates and democracy and the Olympic Games, the returning soldiers flew. Across the Peloponnese

Peninsula and fringing the east coast of that long, glittering finger of the Middle Sea, the Adriatic, the American protectors of the Sinai flew on. Into the ancient, historied heart of Europe the modern-day warriors soared, above field and plain where innumerable ancient warriors had fought their bloody battles, long before there was paper on which to record them. Northwest, to the international airport at Cologne, Germany, where a fresh crew replaced the jetliner's tired handlers. At 1:21 a.m., December 12, her new captain was John Griffin.

With tanks topped up with 45,000 kilograms of European jet fuel, the aircraft lifted off at 3:22 a.m. for the longest leg of her east-to-west run. She flew northwest, past England, following the great circle route west over the North Atlantic. The passengers were jubilant and triumphant, bursting with the joy of partying, meeting family, holding wives and husbands, and the glorious Christmas season was on everyone's mind. There were so many stories to tell, so many more to hear. Sweet mornings of waking next to the ones they loved, the sound of sleepy-eyed children still surprised to see Daddy or Mommy home, the pleasure of simply sleeping in.

Silent, like a winged shadow, the mechanical bird broke out of the night cloud and into a murky sky above a black sea. She came up over the very edge of the Atlantic from the east, her starboard, port, and belly lights flickering as she flew toward the island of Newfoundland.

Captain Griffin stretched, straightened his legs, and came erect and fully alert in his seat. It had been a long flight. No matter how many times he crossed the North Atlantic, he always felt the same. Leaving Gander and flying east out over the broad sea gave him the feeling of leaving home. All captains felt the same. Newfoundland was the last land on this side of the Atlantic, Gander being the last fuel-up before Europe. Flying west again, Newfoundland was the longed-for landfall sought from every cockpit, Gander the welcomed roost for all. Griffin never felt the same way whenever he left or approached the British Isles. Whatever it was about this ragged, triangle-shaped island broaching the very entrance to the North American continent, all airmen breathed easier when it came into view.

Now here it was again, sprawled across the horizon, dark and distinct against the ocean as the jet gained the land. Tiny, flashing lights appeared below the airship. They came from the lighthouses Griffin knew were keeping a constant vigil for other night travellers along this coast of dangers – beacons guiding seafarers to the safety of the island. He counted the delay of one light below them now. One, two, three. The light flickered out. John counted again, one, two, three, and the light was back on again. A three-second light and three-second eclipse. The sky captain knew that this count would aid sea captains in locating the capes and islands where the lighthouses stood.

As he watched, the light winked out of an invisible house on a blackened land, while the night sea coated the island's shores in white.

* * * * *

TOM BRAGG WATCHED AS the lights from a plane came in over the ocean, closing with the land. Tom liked to watch the big airliners approach his island. Some of them flew too high to see much more than a faint, winking light, but others came in low over the ocean. They always provided a break in the monotony of work. A lighthouse keeper's job was often a lonely one. It was the darkest time of night, that pre-dawn hour that always seems to deepen all natural shadow and encourage man's attempt at artificial light. Tom knew all about light and the importance of it. He was one of the lighthouse keepers on Puffin Island.

Puffin Island was only eleven hundred feet long and at its widest barely one thousand feet. The lighthouse stood just eighty-six feet above some of the wildest stretch of North Atlantic Ocean anywhere. The Puffin Island area was on the flyway for airliners travelling the Atlantic route to and from Northern Europe. Gander was in their direct line of flight from this part of the coast. Tom wondered, as he always did, whether or not the pilots knew the lights they were seeing below them were lighthouses. The planes, he reasoned, had their own lighthouses to guide them out of the skies. Control towers, perhaps, but high buildings guiding ships to land from sea or sky were all lighthouses to Tom Bragg.

Puffin Island stood with a quarter-mile stretch of ocean separating it from the island community of Greenspond. The island the French had called Grin d'Espagne had finally been connected to the mainland by a causeway in 1983. The Puffin Island lighthouse had veered seamen away from this northeast section of treacherous Newfoundland coast since 1873. When Tom took the job of keeping the light in 1968, it was just a kerosene lamp the lighthouse keepers lit by hand. In 1985, the revolving spring mechanism that kept the light turning still had to be wound tight three times each night.

The plane Tom had been watching flew low over Puffin Island, the report from its engines loud in this house of stone, and then she was gone in over the land. It was just another plane, like a hundred others the lighthouse keeper had seen before.

* * * * *

CAPTAIN GRIFFIN AND HIS crew made ready to land on Gander's runway. Griffin smiled. They had made it across the Atlantic one more time; they were safe again. A voice from the control tower came through his headset, calm and friendly. He was cleared for landing. The nose of the aircraft tilted down and, dipping like a giant raptor glad to be out of a night sky, the jet roared downward to a gentle landing on Canadian soil.

Most of the troops were fully awake before the plane made land. Many of them peered through the small oval windows along the plane's fuselage at the scattered lights of Gander rushing toward them. It had been a rough night for most of them. Some had slept soundly, slumped in their seats, while others had only dozed for a few semiconscious moments at a time, jolting awake each time their tired heads fell on empty space.

The dreary pre-dawn didn't show any promise of warmth after the plane touched down and parked beside the low terminal building in Gander this December morning. Even the lights of the nearby town had vanished, with only the warm glow from the many terminal windows at the airport illuminating the area. Many of the soldiers left the aircraft without jacket or hat, in spite of the cold. There was no wind at all, but the temperature was near freezing as they walked down the steel steps to the ground and crossed the tarmac to the building. For soldiers who had just spent six months in a desert climate, it felt like the North Pole.

Once inside, all traces of cold and night vanished from the minds of the Americans. Bright lights and friendly people met them. The staff at Gander Airport were in a Christmas mood, greeting them with cheerful Merry Christmases. Festive music seemed to come from everywhere. A tastefully decorated tree stood in the centre of the terminal. For the homebound soldiers, the Christmas season had begun.

* * * * *

CYNTHIA GOODYEAR WATCHED THE soldiers as they walked up through "The Finger" – the narrow corridor leading from the tarmac to the international section of the terminal. She knew that once inside the building, the American soldiers were free to roam anywhere they pleased. She also knew they would all walk through her shop. The soldiers had been expected by the Gander staff, their turnaround part of the routine of this international airport on the very edge of the North Atlantic. Arrow Air was affectionately known to all at the airport as the "Big A."

Cynthia Abbott was born on December 20, 1948, in the tiny fishing outport of Ragged Harbour, which would in time be little more than an hour's drive from Gander. On the cold December day when Cynthia was born, the only avenue out of the community was the winter sea. Ragged Harbour was a harsh place for fishermen, situated on the northeast coast with its long, straight shore wide open to a relentless Atlantic. But clear of the land and out to sea swam the livelihood for almost all of its residents. It was a place of rugged beauty, with its long, pure white beaches sloping down to the rolling sea.

By the time Cynthia was in her teens, her town had lost its wonderful and appropriate "Ragged" name to the nearby and larger Musgrave Harbour, named in 1886 for its English governor, Anthony Musgrave. The Musgrave Harbour name was by this time known across the country and beyond, forever linked with Gander and aviation history and the death of one man.

Doctor Frederick Banting was the sole passenger aboard a Hudson Bomber that left the high central plateau of the airport at Gander on the night of February 20, 1941. They were headed for England. The bomber, designed to drop screaming death on the bloodied plains of Europe, carried this night the man who would be responsible for saving the lives of millions of diabetes victims the world over. Banting had been awarded the Nobel Prize in medicine in 1926 – a prize he shared with his co-discoverer, Charles Best. Banting himself had no idea the impact his discovery would have. Nor would he ever know. Before this night was over, Doctor Frederick Banting would die.

Barely settled into their night flight less than fifty miles east of Musgrave Harbour's bleak coast, over a stormy, black Atlantic, the bomber ran into serious engine trouble. Captain Joseph Mackay and his co-pilot, William Snailham, decided to turn back to Gander. Their navigator, William Bird, gave them a hurried course back to the airport. They would not make it. Sputtering in from the sea, the Hudson flew over Musgrave Harbour. It fell from the sky and slammed into a frozen, tree-lined bog just a few miles inland, taking the life of its co-handler and its navigator. The man who was knighted by King George himself in 1934 – Sir Frederick Grant Banting – died with them. The thirty-three-year-old captain survived.

Cynthia came to Gander in search of work in 1967. The dusty, muddy, pothole-filled new road had opened new opportunities for the industrious young woman. It didn't take long for the attractive brunette with her easygoing, shy, outport manner to catch the attention of the young men. She met Wilson Goodyear, a heavy-equipment operator. They fell in love and were married in 1969.

Cynthia worked in the duty-free shop owned by the Cara Operations Company at the Gander terminal. She loved her job. Everyone came into her shop, including the international travellers. Cynthia loved meeting new people. She had made friends all over the world. This morning she had been told the soldiers from Arrow Air would be arriving soon. She looked forward to their visit; the American soldiers were always so polite. Cynthia especially loved the ones with the southern accents. She wondered if she would recognize any of them from six months ago, when they had flown east. They had spent six months in the Sinai and now were going home.

Just in time for Christmas, she thought. *There will be many celebrations in the States tonight.*

Many of the soldiers walked past the Big Dipper bar. Like everything in the town of Gander, the bar had an aerial connection to its name. The lounge took its name from that most magnificent of constellations, the Great Bear, with its easily recognizable drinking gourd visible in every starry northern sky. The

Big Dipper had once been the largest airport lounge in the world.

Even at that hour some of the American soldiers felt like they could use a cold Canadian brand of beer. The bar was not open, nor would it be to them. Their commanding officer had a strict rule about his travelling soldiers, simple and direct: there would be no consuming of alcohol. Any alcohol purchased in the duty-free shop was not allowed on their person or in their carry-on luggage.

Inside the duty-free store they bought Christmas cards and small Newfoundland flags, trinkets, and fridge magnets. The most popular item was the T-shirt with I SURVIVED GANDER NEWFOUNDLAND printed across the chest. One soldier put his shirt on in the shop. Every one of the soldiers bought something. Cynthia wasn't surprised when not one bottle of booze was purchased. Not even one bottle of "Newfie Screech" came to her counter. They bought no cigarettes, but children's toys and books instead. One of the female soldiers bought a pack of scented candles, another a postcard with a magnificent iceberg rising out of a blue sea. When American soldiers stopped here on their way east they rarely visited Cynthia's shop, but the homebound troops always did. Cynthia was pleased to see they were all in a Christmas mood, their mirth and goodwill buoying her own pre-Christmas spirit.

One short young man with sparkling white teeth walked to the counter. He laid his purchase down and Cynthia looked from his sun-browned face to see what he had bought. Both his big hands were splayed on the counter. On the middle finger of his right hand was the biggest, most ornate ring Cynthia had ever seen. The wide band was gold, crowned, and encased with what looked like large diamonds and glistening emeralds. Tiny red and green garnets peeked and shone from the jewel. Cynthia had seen many fancy pieces of jewellery worn by world travellers, but never had she seen anything like the one the young soldier sported on his finger. She reached for a bag to wrap the items the man had bought.

"My goodness, Marie, did you ever see such a ring?" Cynthia whispered.

Marie Lawton worked the morning shift with Cynthia. She turned and looked first at the man's face – he was very handsome – and then at the ring.

"It must be worth a fortune. I never did see the like. And on a soldier's finger, too."

Cynthia placed the man's item in the bag. He pulled an American twenty from a leather wallet and offered the bill to her with his right hand, the ring boldly displayed.

Cynthia couldn't resist. "I couldn't help but notice your ring, sir. It's unlike anything I have ever seen. It is striking."

The soldier removed his change from the counter, looked at Cynthia with piercing grey eyes, and, without looking at the ring or acknowledging her compliment, walked away.

* * * * *

CHRISTMAS MUSIC DRIFTED SOFTLY from the speakers. It made the travellers feel right at home. Some of the soldiers smoked American cigarettes, others the stronger Middle East kind. Darrin Brady wasn't a smoker. He stared at the wall-length painting in the terminal building, his eye drawn instinctively to the birds of flight. It reminded him of the drawing in his pocket. Darrin envied the artist's ability to create such an image. He took the drawing from his pocket and examined it again. Michael still wanted the sketch, but Darrin decided he would keep it and maybe submit it to an art contest when he got home. He would show it to Shelley first and get her opinion. She always gave him an honest answer.

The Christmas instrumental played an age-old tune. The upbeat feeling at the terminal awakened the soldiers from the lethargic doldrums of the long Atlantic crossing. Almost all of the troops looked at Gander as being a second home. The history of this island and their own country was forever linked. Before this day aged, the 101st Airborne Division would make that link indelible.

In the duty-free shop Darrin bought an I SURVIVED GANDER T-shirt. Several of the others bought one as well. A few bought comic books to read on the last leg of their trip. Still others bought snack food. Darrin walked to the payphone area, thinking he should call home. While waiting in line he overheard one of his fellow soldiers say into the phone, "Tell her Daddy will be there before her bedtime. I am almost home. I love you . . . Merry Christmas. See you tonight." Darrin smiled at the soldier and stepped to the phone. He looked at his watch and realized it needed adjusting for yet another time zone.

The pre-boarding announcement sounded through the terminal. Darrin figured it was too early to call his mom and dad anyway. They would still be asleep. He thought about Sue. A man should always have a sweetheart to come home to, he finally decided. Maybe he and Sue would settle their differences.

Darrin walked away from the phones and studied the mural

again until the final call to board came over the speakers. He wondered if he would be going back for another tour in the Sinai. With the Army, you never knew. Darrin still didn't know if he wanted to stay in the Forces. He would make that decision during his Christmas break. Falling in line to walk down the ramp for their final leg, Michael walked by his side.

"The best part of the trip, eh, Buckeye? The last leg is always the best one."

Before Darrin could answer, a voice sang out behind them. "One more river to cross."

Darrin turned and laughed at a smiling Shayne. "Don't you mean 'ocean,' Shayne?"

"No matter, man, it's the last water before home. Salt or fresh, makes no difference to me." Shayne expanded into another rough tune, grinning as he did. "Cause my baby's on the other side." Everyone who heard him laughed along with him. It was indeed the best part of any long trip. The 101st were almost home.

CHAPTER 4

Today, Robyn Stack lives in Athens, Texas. She had recently reached the age of seventy and really felt as good as she had when she was sixty. It was partly due to her excellent genes, she figured. Robyn shared a home with her very active and fiercely independent ninety-seven-year-old mother, Kaye Blew.

Robyn had spent a good part of her working years as a professional publisher. Now retired, she was amazed at how remarkable a tool the computer had become. It could be a great source of entertainment, an inexhaustible source of information, and, as she had just found out, it could revive old memories.

Earlier today she had been looking through her email. She and her mother had just returned from a hairdresser's appointment – her mother made faithful visits to keep her full head of hair preened just right. There was only one email. The subject field read RE: GANDER ARROW AIR CRASH.

Robyn sat back from her desk. The small screen with the four glaring words jolted her. Her stomach was in knots as her mind dragged the buried memories back, as it always did at the mention of the name Gander. She opened the file.

It was from an author who lived near Gander, Newfoundland. He wanted to know if she was the same Robyn Stack who had lost a son, Shayne Stack, in the Arrow Air crash on December 12, 1985. He was writing a book about it, the note read, and would she be willing to share her son's story with him?

Robyn leaned away from the computer keyboard again. What could she say to this man? Maybe she wouldn't answer at all. He was probably just another one who figured he had all the answers to the Arrow Air crash. She pulled her chair closer to the computer and finished reading the email. This guy stated clearly that he was only concerned about the human interest side of the crash. He would write absolutely nothing about the controversy surrounding the disaster. This was different, she thought, and interesting. The author also provided a link to a website where she could learn about him, to make sure he was authentic.

Two days later Robyn opened the email again and clicked REPLY. She would disclose basic information about her son, nothing more. Robyn didn't realize then how fully this unknown man from Canada would awaken memories of her son that she thought she had filed away forever. Once she started, she couldn't stop. Putting all her thoughts and memories into an email was altogether different from talking about them. The author had unintentionally opened for her a healing process of sorts. Robyn told her boy's story by degrees, as if by doing so she could finally put him to rest.

* * * * *

WHERE EAGLES LIE FALLEN

SHAYNE STACK WAS IN pain. He could never understand why airliners had to fly so high. He knew turbulence from changing air currents was less likely the higher they climbed, but he wondered if the same calm air couldn't be found a little below 30,000 feet, which seemed to be his pain threshold. High altitudes always caused him pain in the mouth. It wasn't anything he couldn't bear. He rarely spoke about it, but it happened every time he flew. It was June 1985 and Shayne was headed to Egypt as part of the Screaming Eagles contingent serving with the MFO in their Sinai south camp.

Shayne was one of the proudest members of the group, the fierce fighting history of what he now called his battalion a huge motivator for his restless spirit. He had signed up for the Army in 1982 and got through Fort Benning boot camp in Georgia. Basic training was not his favourite part of the infantry. Shayne was a let's-get-on-with-it kind of guy. After basic training he was assigned to the 1/87th infantry – B Company – and in 1984 he served two tours in Germany before finally becoming part of the 101st Airborne Division in Fort Campbell, Kentucky.

The decision to get into the Armed Forces had been a fairly easy one for Shayne. Both of his sisters had already joined up. His older sister, Stormy, had joined back in 1979. Shayne loved her name. After breezing through basic training, Stormy had been assigned to the American base in Germany. His younger sister, Teresa, finished high school in 1980 and tried the University of Columbia for a few months before deciding it wasn't for her and left. Terry joined the Air Force and was sent to Guam, the southernmost island of the Marianas in the Western Pacific and a territory of the United States. Terry always boasted as being "where America's day begins."

Shayne had a brother, Bill, who lived with his father in Hong Kong. To the restless Shayne they were all seeing the world, getting the job done. He had no intention of settling for any nine-to-five job. True to his fighting nature, he set his sights on the elite of America's warriors, the 101st.

Shayne gripped his nostrils and blew out softly, equalizing the pressure inside his head. The slight discomfort that almost everyone experienced from high-altitude flying wasn't a problem for him; as a diver, Shayne knew how to relieve the pressure in his head. It was the sharp pain in the roof of his mouth and both his upper and lower jaws that gave him so much misery. The source of Shayne's high-flying pain had come from a baseball game.

* * * * *

MICHAEL SHAYNE STACK WAS born on February 7, 1961, in St. Louis, Missouri. His parents called him Shayne. The Missouri River used to be the longest river in America, but a few miles of river channelling narrowed its length to that of the Mississippi. It was near the confluence of these mighty rivers a young Shayne grew and spent his boisterous boyhood. From the beginning Shayne was not an ordinary child. Nothing he did and nothing that happened to the boy were ever in half measures.

Shayne developed allergies that progressed into asthma severe enough that he had to undergo a battery of painful injections for the first three years of his life. None of it slowed the boy down; his energy seemed boundless. His parents moved from Missouri to the southern part of sunny California in 1964. The climate change seemed to be a cure for Shayne's ailments.

Before the year was out, his asthma had all but disappeared. However, it was replaced by something far worse. A bad case of chicken pox had worked its way into both his eyes.

He developed the mumps with painful swelling of the salivary glands, and in Shayne's case the parotid gland. It was accompanied by testicular swelling, which doctors said could cause sterility. The virus came with a burning, itching rash. The mumps virus is a form of meningitis sometimes accompanied by a severe rash. In a matter of days, Shayne was diagnosed with meningitis.

Robyn, his mom, was terrified. Just to hear the doctor say "your son" and "meningitis" in the same sentence frightened her. Robyn knew a little about the disease. It was an inflammation of the protective membrane that covered the brain and spinal cord.

Shayne went through bouts of increased heart rate and reduced blood pressure. His body went from extreme highs in temperature to extreme lows. His breathing became rapid. The doctors feared blood clotting, which could prevent adequate blood flow needed to keep the boy alive. They were very concerned and told Shayne's mother about the possibility of hemorrhaging of the adrenal gland. It was a death sentence for many children.

However, Shayne survived it all, unscathed. His youthful will, combined with his always positive attitude, carried the boy through. By 1965 Shayne's family had moved again, away from the coast and inland, this time to Arkansas, a state named for another river that pours its issue into the Mississippi.

Shayne was fearless, game for anything. There wasn't anything the eager boy wouldn't try – even self-propelled flight. At the age of five, a very determined Shayne decided he was going to fly, on his own power. If Superman could do it, so could he. His flying debut was not a summer day's whim. It was a premeditated and well-conceived idea. Shayne figured if the kids wanted to see him fly they should pay, so in true carnival spirit, he charged the neighbourhood kids to see the event. They had to have a ticket to enter his yard.

He prepared a ladder, and Shayne, with a long, dark cape draped around his neck – looking more like a modern-day Zorro than Superman – climbed to the very top, gave a Geronimo yell, and jumped into the air. The wild cry when he jumped didn't bring his mother – she was used to it – but the scream of pain when her son crashed to the earth did. Shayne had broken his arm. It only slowed him for a while; he was very proud of his "mummy arm."

Before the cast had done its healing work, Shayne had broken his arm again – the same one. The cast had to be removed and his arm broken yet again, in order for the doctors to reset it properly. Shayne had a brand new "mummy arm" to sport around. He was the neighbourhood hero, of the boys and especially the girls. Shayne's next episode with pain would not be planned, and with a cruel twist from fate's fickle hand would cause even more pain far into the future – after his death.

Shayne's parents were divorced by 1970, when Shayne was nine years old. Robyn Stack was suddenly a single mom with four small children. She had two daughters. Stormy was a year older than Shayne. Terry was just a year younger. Shayne's brother, Bill, was four.

If there was ever a child who needed a male role model, it was Shayne. When he was thirteen he went to live with his father. It helped.

Shayne loved sports, especially baseball. He wanted to join a little league team. His mother agreed it would be an excellent place for Shayne to spend his limitless energy. Before long he was chosen to play catcher. Shayne was a fast, fearless, and very aggressive player. He always played to win.

One day they were playing in a deserted field on the outskirts of town. There were only a few bleachers, with no cheering crowds and only proud parents, relatives, and friends watching and rooting for their favourites. Shayne, the catcher, was taking his turn at bat. The pitcher grinned at him, intending to intimidate him. It didn't work. The pitcher circled his mound and waited, trying to get his opponent to relax his guard. That didn't work either. Suddenly he pitched, and all of the boy's young muscle poured into the throw. It was a curveball. Shayne placed his bat back over his shoulder, judged the rushing ball just right, stepped into its arc, and swung in one easy motion. The crack of the bat against the baseball was music to his ears.

For an instant Robyn could see her son's white teeth flash in his trademark devillish grin. The ball flew high up into the air, what looked to be a fly ball, but the outfielder fumbled it. Shayne knew it wasn't a good hit. He tossed his bat aside and raced for second base, determined to beat the throw. The second baseman yelled to the outfielder. More than the length of his body away from second base, Shayne knew he wasn't going to make it, at least not on his feet. He decided to slide. His mother saw him lunge toward the dusty ground and knew his intent. *Another ripped shirt to mend,* she thought.

Shayne met the ground supine and slid like a luger, his arms stretched out for the plate. The outfielder had thrown a bad grounder; the second baseman would never reach it in time. Suddenly there was a sound like a hard-flung ball making contact with a determined bat. Everyone in the bleachers except Robyn Stack was straining to see what had happened. A cry came from the prone boy at second base, a wail of agony and helplessness. His mother was already racing across the field.

The second baseman was walking around and around the screaming boy, not knowing what to do. Blood mixed with saliva poured out of Shayne's destroyed mouth. The dusty ball was streaked with his blood. The boy's lips were torn and his jaws didn't look right. Shayne's bloodied hand held some of his teeth and several more were on the ground. Robyn grabbed her boy, desperation giving her the strength to carry him. She yelled behind for someone to pick up his teeth as she put him in the car. She tried not to keep staring at the mangled mess that was her son's face. Shayne's neck was splayed with his blood and his shirt was soaked with it, but he wasn't bleeding much. Robyn thanked God for that and sped toward the city. Shayne held his jaws with both bloody hands and sobbed quietly. Even though the boy was crying, the pain unbearable, he showed no fear. A policeman trailed Robyn's speeding car, saw Shayne's bloody face, and escorted them to the hospital.

The baseball had torn from the boy's tender mouth one tooth for each of his twelve years. The bone at the roof of his mouth was shattered and would require a splint. His broken jaws had to be wired closed; he wouldn't be eating solid foods for a long time. He wouldn't be talking, either. Miraculously, the doctors managed to place all of his teeth back into the broken gums. For months Shayne sucked his puréed food through a straw. He never complained.

He even found ways to make light of his situation. One day they were having pizza. His mother liked a glass of beer with hers. Shayne placed a slice of the tasty pie in the blender, waited for his mother to turn away, grabbed her beer, poured half of it into the blender, and before she could stop him he had sucked most of it down, his eyes gleaming with mischief.

One year passed before the splint was removed, the wires pulled from his jaws, and his mouth had finally healed. Shayne wanted to play ball again. He sauntered out onto the field as if he had been playing all the while, his catcher's glove swinging as he went. He crouched behind home plate and waited for the throw. Robyn sat in the bleachers, her hands covering her worried face. She was praying in silence, sure that Shayne was going to be dealt another terrible blow. She was close enough to hear the splat of the ball as it hit leather. Looking through her opened fingers, she saw that Shayne was facing her, one gloved hand holding the ball, his other hand thumbing high in victory. On his face was the broadest, whitest smile his mother had ever seen. Robyn would never really worry about Shayne again. He was a survivor.

* * * * *

THE THROBBING PAIN IN his jaws began to fade away. Shayne knew they were descending into Cologne, Germany, for refuelling. Rubbing his lower jaw, he smiled at the memory of that long-ago game. He still loved a good game of baseball, and at one inch past six feet and weighing just a bit over one hundred and eighty pounds, he could be a formidable player. He still preferred to play catcher.

Germany was no stranger to Shayne. He had travelled here several times before. In 1983 he had first been stationed in Baumholder, in southwestern Germany. The following year he had been assigned to Mainz, part of the fertile European Rhineland. The city of Mainz was on the west bank of the Rhine River, opposite the merging river, Main, which added to its northbound flow. Mainz, with its more than two thousand years of permanent settled history, was once conquered by the Holy Roman Empire and became the seat for the Roman Prince-elector. The Roman general Drusus established a command fort there as early as 13 BC. Mainz had been under foreign control for years, its strategic place in the great bend of the Rhine, with its North Sea mouth open to the broad Atlantic, forever coveted by would-be conquerors.

Mainz went from Frankish rule in the fourth century to the Christians in the Middle Ages. The tenth century saw the Jews in control, the French Revolutionary Army in 1792, the Hessians in 1816, and the German Empire in 1871. It fell to the French again in 1919. Under Adolf Hitler, who became Chancellor in January 1933, Mainz was again under German control. It was next fought over, and finally fell, to General George Smith Patton of the Third Army – 90th Division, United States Army – on March 22, 1945. After the close of World War II, Mainz again fell under French control, as part of the Allied agreement until 1949, when France withdrew her presence. Now in 1985, Mainz had another military presence. The USAREUR, or U.S. Army Europe, occupied military bases out of the McCully Barracks in Wackernheim and the Mainz Sand Dunes. They were used for training purposes only. Mainz was once again under its own governance.

Shayne had many fond memories of his assigned tours in

Germany. He loved the country. He liked the food. He loved the beers always served in generous, frothing steins. Shayne wasn't a light-beer drinker. He especially loved the fair-haired, fun-loving women. It was while in Germany that Shayne had bonded with the most special of all women in his life – his mother.

Robyn Stack worked with a very successful publishing company. She was frequently sent to many countries to represent her publisher. One of the book fairs she was asked to attend was in Frankfurt, Germany. Shayne arranged furlough, she booked rail passage, and mother and son toured Europe. Their twin Nikon cameras captured the majesty of Switzerland's alps, sunsets in Paris, the City of Light, and Venice with its twisting watery avenues. Somewhere between the serenity of the Alpine Mountains and the blue of the Adriatic Sea, mother and son found, without really seeking, a rare gift of love and friendship. They would never be truly apart again.

Watching the land rise toward him, Shayne smiled to himself. He had another memory of his days in Germany. Most of the servicemen had nicknames applied to them. Shayne shared his surname with Robert Stack of the popular TV show *The Untouchables*. Robert Stack starred as the intrepid crime fighter Eliot Ness, and so Shayne was nicknamed "Eliot." He liked it. While in Germany he had the name tattooed on his left thigh. Shayne rubbed his leg, still remembering the burning sensation he had felt for days after the name had been forever inscribed into his flesh.

CHAPTER 5

THEY WERE WELL INTO their descent for Cologne. The plane dropped into a small air pocket and lurched a bit as it met the air currents rising from the plains of central Europe. The sudden motion jolted Jeff Kee awake. He was in the aisle seat next to Shayne.

"Hey, Keybird, that last pothole finally got you awake, I see." Everyone called Jeff "Keybird."

Jeff grinned his easy smile, rubbed his eyes, glanced out the window, tightened his seat belt, and said, "You know, Eliot, I was hoping to wake up next to someone a lot prettier than you."

"When a man wakes with a smile on his face, he had to be dreamin' of a woman. Was it Tracey, or are you already cheating in your dreams?"

"Oh, it's Tracey for me, all right, forever. What about you and Jill? You two didn't seem to be hittin' it off real good when we left stateside," replied a concerned Jeff.

"Aw, I don't know, man. Jill and I have been together forever. You know, high school and stuff. She's one of my best friends – a female one, anyway. Well, besides my sister. She always understands me. I was engaged to be married, once. A girl from Texas – Tammy. It didn't work for us. So now Jill and I will take it slow, you know, make sure we are right for each other."

Shayne left it at that, not wanting to share everything in his personal life with his soldier friends, not even Keybird.

Shayne still remembered the sad look in his mom's eyes when he had told her he and Tammy were not to be married. It was just a couple of weeks before he shipped out for Egypt. She had said little about it. She rarely interfered with his decisions.

Jill was part of Shayne's crowd and the best friend of his sister Terry. She came for visits to his mother's house. Everyone loved the dark-haired young woman. She had driven Shayne to the Fort Campbell base for his flight to Sinai.

Keybird, Tracey, Shayne, and Jill had spent that last night drinking beer and partying hard. The next morning the four friends shared goodbyes. Shayne noticed the effort it had taken for Jeff to get the sobbing Tracey out of his arms. He had walked away, not wanting to see a soldier cry.

Shayne had held Jill tight. Her misted eyes were confused as she wondered where their love was headed. Shayne had no answers then, and none now.

Shayne had returned from Germany in January of 1985 and was transferred from B Company 1/87 Infantry to the 101st Airborne at Fort Campbell. He was pleased. The 101st suited him fine. He figured being part of a fast-paced, high-energy group was the place for him. He couldn't wait to "hit the silk" from a soaring plane and become part of the "stick" of such a famous group of sky troopers. Shayne excelled at it. Fearless, he was eager to try the jumps; the adrenaline rush of free falling before he pulled the ripcord always filled him with excitement. Each time Shayne jumped his mind flashed back to the day when he thought he could fly from the top of the ladder. Now he could yell Geronimo in earnest; they all did it. Shayne's yell always seemed to be the loudest.

Shayne was a lover of all fast-moving machines, and in true Shayne Stack fashion, his hectic lifestyle almost cost him dearly many times. He had lost control of his Harley on the Dallas Freeway once and flipped the powerful bike like a circus act gone bad. Even as he was propelling through the air, he knew his bike was history. Shayne walked away from the crash with a broken collarbone. The Fort Campbell parachuting and rappelling from helicopters always aggravated his neck injury. He never told anyone about his pain.

He was assigned for duty as part of the MFO in Egypt's Sinai Desert. Shayne was about to learn that peacekeeping missions could be very dangerous. They could also be that dreaded of all things – boring.

The first thing that impressed Shayne in Egypt were the endless sand dunes. *God, what a place for a dune buggy*, he thought. He loved the feel of a warm wind rushing through his hair as he drove up and over the sand. He had even designed and built a dune buggy of his own one time. Shayne had always been good with his hands.

The second thing Shayne noticed about the Sinai made an even bigger impression on him. It was the terrible heat. He, like all the others, had been warned about this. He didn't like it at all. Shayne had figured it was an exaggeration – after all, soldiers always told rookies that things were worse than they actually were. However, this time they were right. He had never felt such heat. Even Southern California and the Texas Plains were cool by comparison. Shayne did not do well in high temperatures, and as he was about to find out, the Sinai could be extreme. What he found even more astonishing was how some of the nights could be so cold. When the sun slid down in the west and took the intense heat from the land, the soft, purple twilight crept over the earth with a frigid night wind.

Shayne had been part of a family outing in the Westport area of St. Louis when he was eleven years old. The fairgrounds were crowded, noisy, and very hot. He became so overheated he fell to the dusty ground, and while the watching crowd made room for the fallen boy, he went into convulsions. He was rushed to hospital and given a complete examination, even an encephalogram on his brain. He checked out fine.

One summer during his high school years, Shayne had taken a job with an arboriculture company. Part of his work involved tree trimming. One hot, dry day without a draft of wind, Shayne was high in a tree lopping off branches. He became so overcome with the heat he fell from the tree. He slid down from branch to branch, unconscious before he hit the ground. Luckily, his body's limp state had protected him from serious harm. He was diagnosed with heat stroke again. Now here he was in one of the hottest, driest places in the world. Shayne didn't like it.

The heat of the Sinai summer wore on. The endless days of sandbagging, rotating guard duty night and day, along with endless patrolling of the same dusty roads with the constant fear of carefully hidden landmines, took their toll on him. Shayne found he couldn't be his old self. His pent-up energies went unsatisfied. He shared his frustrations with his mother.

7/27/85 — Sinai

Hi Mom:
Things over here are pretty hot, yes indeed. Well, I have been out in the desert for ten days now and twenty-four more days to go. The sun comes up at 0500 and the temperature is about 85°F out. We do our P.T., eat breakfast, take showers. By 0800 we do our morning details and classes. By 10:30 or eleven it is 115–120°F. One day it got up to 136°.

We have a huge minefield right up the road from us that is 10 miles square. Pretty bleak situation. Most of our time is spent filling sandbags and pulling guard duty. The desert over here is a dump. All around our position there are signs of war: old bunkers that are blown up, parts of vehicles, minefields. Even when you get into a town there are more police than people.

We did get to spend a couple of days in South Camp before we came out here. I got to do a few things like snorkelling around the reefs. We jumped the fence to get out of the place and went to his beach about 5 miles away to go snorkelling. It was picture-perfect. I have never seen so many colors in fish before. This is supposed to be the 2nd best place in the world for scuba diving.

We got caught coming back on post and it cost us $10 apiece to bribe the guard, but it was worth it. We are supposed to have 10 days off next month, around the 15th or so. That is when we might get to do some traveling. I will write back when some of the pictures are developed. Please send everybody's address. I left it all at home.

Love,
Shayne

Shayne got his days off. He and a friend travelled around Egypt. He didn't like the local beer much, but drank it anyway. He noticed how some of the girls – and a few of the older women – looked at him as he passed, their black eyes friendly and strangely inviting. It angered him to see some of the same women bend their heads, with eyes downcast, when they saw men from their own land approach.

Shayne, who always suffered from the heat, wondered how the girls could endure it underneath their flowing garments. Only a few wore what appeared to be silk dresses. Most were garbed in robes of a dreary black material that had to be stifling. Though he said hello to many as they passed, none of them answered him.

He rode a smelly Arabian camel with one hump to the Giza pyramids. Shayne thought the one-humped variety of camel was called a dromedary, but he wasn't sure. He made a mental note to ask Michael. The camel had grey spots on its curved neck and slobbered long strings of saliva from its constantly sideways-moving jaws. It also appeared to blink in perfect sync with the movement of its jaws. They stayed around these mortuaries of the ancient pharaohs until night came. Shayne's "dromedary" wouldn't kneel for him to dismount. After he had slithered down over the animal's odorous rump, it promptly folded all four legs beneath its rounded belly, closed its hairy eyelids over big dark eyes, and appeared to fall instantly asleep, the animal's saliva still dripping from its grinding jaws.

Standing below one of the world's greatest mysteries and watching the night come to the desert, Shayne was as much in awe of the structure as the millions who had come before him had been. His first thought – in classic Shayne fashion – was that these pharaohs must have had enormous egos. To build such a thing out here in this barren desert, with stone hauled from God knew where, to build such a colossal tomb to hide their bones, was unthinkable. And to use slave labourers, many of whom died to erect these monuments . . . Shayne shook his head in amazement.

A wind came with the gathering darkness, bringing with it the warm, rich smells of an ancient Eastern land. The *khamsin* sighed and whispered around the raised monument as if it were rustling through the green leaves of a majestic cedar. Shayne stood back to better see this wonder. A sheen of stars circled the pyramid's peak. Finally he reasoned it didn't matter at all why the pyramids were here. They were here, for all to marvel at. It was enough. The soldier was silent.

Maybe this is why we are here in this place, he thought. *So that man will always be free to build his pyramids.*

Summer melted into autumn and still the terrible heat continued. Part of Shayne's battalion was assigned to a high limestone outcrop. The hill rose up out of the Sinai like a gigantic anthill. At the very top was a lookout tower fortified with sandbags and cut fuel drums filled with sand. Despite its height above the desert, it provided no cool draft. The infrequent wind that did come was a heated one, borne from the depths of one of the hottest places on earth. Shayne promptly named this corner of the world "The Rock."

8/27/85

Hey Mom:

How are you? I got your letter today and we finally get off this "Rock" tomorrow. Can't wait. We do have a couple of pets out here: 1 dog, 3 scorpions, a lizard, and a snake. We have a guy here that thinks anything that moves is a new pet. He walks around for hours lifting up rocks looking for a new pet. Not much to do out here except kill flies.

Do you know that in one hour during the day, sitting in the guard shack, you can kill over 50 flies? We have a bucket half filled with dead flies.

Oh yes, those sandbags. They surround our building so if anybody shoots at us they will hopefully slow down or maybe stop the bullets, if they do what they are supposed to do. I will send some pictures to explain better in the next letter.

See ya in December.

Love,
Shayne

Shayne's shift on "The Rock" was over the next day. He returned to camp in the same heated desert, suffering in silence.

* * * * *

WHILE ON A WEEKEND pass to get away for a much-needed change of venue, Shayne and two of his buddies met a couple of girls who were on vacation in Egypt. One of them had a German accent. She pronounced his name *Sean*, the German way. Shayne suggested smuggling them into the back of his barracks at Sharm el-Sheikh and the girls agreed. It was tricky getting them past the guards, but Shayne had his ways.

Once inside the barracks, Shayne and the girl with the German accent found a common connection. She was from Stuttgart, Germany. Shayne knew the place from his days stationed in that country. The soldiers and the civilian girls talked about sports and holidays, movie stars and music, about home, about war and terrorism and the mess the world was in. Keybird told everyone that he and Tracey would be married this very month, on December 16, just a few days after he got home. It would be an early Christmas gift, he said, a celebration that would continue into the Christmas season. Shayne could see the love in his friend's eyes. He envied him but said nothing.

The gathered young people even discussed the Bible. It was an easy topic to get into, here in the land where most of the great Bible adventure stories had begun. Several more of the soldiers joined them. It was an easy, pleasant time spent among young, modern friends. Where they were all from, their mixed backgrounds, their skin colours, their beliefs, all faded into a simple evening of laughter and sharing of ideas, a lesson the leaders in these troubled parts would have done well to learn. The next day they parted, the German girl and her companions continuing with their travel plans. The soldiers prepared to leave the Sinai. It was time to go home.

* * * * *

WALKING ACROSS THE TARMAC toward the brightly lit Gander terminal, Shayne Stack was glad of the cold. He wore no jacket, nor did he complain of the drastic drop in temperature. He did not like the heat of the Sinai, and he wasn't planning on returning.

Shayne was in the middle of the line. When the point soldier reached the door, it was opened for him. Instantly the sweet strains of Christmas melodies sounded through the doorway, welcoming the Eagles onto friendly Canadian soil.

Inside, Shayne became his old boisterous self. He twirled on the floor to the Christmas music. Several of the others did the same. Shayne teased Jeff Kee about not stepping out on the floor for a mock dance.

"God knows you need the practise. Your wedding day is only a few days away!"

"The one I'll be dancing with on my wedding day is a better dancer than you, and a whole lot nicer-looking," said Keybird.

"Yeah, and she has better-looking legs, too," Shayne said with a laugh.

Inside the duty-free store he picked up one of the I SURVIVED GANDER NEWFOUNDLAND shirts, for his grandmother. He knew she would wear it. A handcrafted lobster pot no more than two inches long caught his eye.

"It would make an excellent tree ornament, my love," one of the ladies in the shop told him. Shayne loved her friendly way of talking. He laid his gifts on the counter, looked full into the dark eyes of the attractive brunette, and for a second remembered another woman he had met during his stay in the Middle East, who had looked away when their eyes had met.

Smiling his widest smile, Shayne said, "You're right, ma'am. I'll take it."

* * * * *

SERGEANT JEFF KEE LOOKED around the small shop. He already had several gifts for his bride-to-be, one he had picked up in the Eastern lands that he couldn't wait to see her wear. The shop displayed many curios, most of which he looked through without interest. There was one display case that drew his attention. Behind the glass were dozens of miniature carved wooden maps of the island of Newfoundland hanging from gold and silver chains. On one side of the maps were etched given names. Jeff searched for the name Tracey, but it wasn't there. There was one spelled T-R-A-C-Y, without the "E." He asked the sales lady if they did engraving.

"No, my love, we don't," she said with genuine regret. Jeff grinned. This was the only place he knew in the world where strangers called you "my love."

He looked at the I SURVIVED GANDER NEWFOUNDLAND T-shirts and decided not to buy one. If the shirt had read "I survived the Sinai," he would have had one. He bought a beautiful pewter box to hold two rings, paid the smiling lady, and told her to keep the change before hurrying out the door. Back in the spacious international section of the terminal, many of the soldiers were singing along with the Christmas tunes. Jeff showed Shayne the ring box.

"That's so cool, Keybird. I envy you, man. You and Tracey really have it all figured out. Hey, soon she'll be called Mrs. Keybird."

Jeff smiled. "You know, Eliot, I kinda thought you and Jill had it together, too. That last night in Kentucky you guys seemed to be hitting it off."

Shayne stared at Keybird for a long time before answering. "Hey! Man, I forgot to tell you about that. Jill is going to meet me when I get to Fort Campbell. We've made no real commitments, but who knows? I have a good feeling about it. You know what? I really miss Jill."

Jeff gave his friend a high-five before crossing the terminal to the men's washroom. A boarding announcement for Arrow Air interrupted the Christmas music. Jeff felt uneasy again. It was the same gut feeling of something just not being right he had experienced before boarding the plane in Cairo, and again in Germany. Jeff was afraid to fly Arrow Air. He had felt so strongly about this, he had told Tracey about it in his last letter home. It wasn't something he could explain, or even tell anyone else. He just had a feeling that this flight wasn't going to get him home.

Shayne and Jeff walked side by side across the wide terminal. A friendly "Merry Christmas" came from the two women at the door of the duty-free shop. Many of the soldiers shouted the greeting back at them. Shayne called "Merry Christmas," and followed with a playful, "my love." He couldn't resist. Flashing his mysterious grin, one his mother had always called his James Dean smile, he fell in line with the others. Shayne and Jeff headed for the finger. It was time.

* * * * *

CATHY ZIEGLER WAS BORN on May 11, 1964, in Berkley, Michigan, and given the name Cathleen Marie Ziegler. From that small town just north of Detroit her family moved to Novi, Michigan, in 1972. Cathy was not an A student in high school, but with hard work and determination she made it through just fine. She loved animals, especially horses. Fresh out of school, Cathy went directly into veterinary classes at Michigan State University. The young girl spent only one semester at MSU before leaving school for no specific reason. Cathy just seemed to want to get on with her life.

Her love for animals stayed with her, and just over a year later she applied for a veterinary position in the American Army. There were no positions available, so she decided to try her hand at mechanics instead. She enlisted in the Army in 1983, into the small engine and vehicle repair program. She was soon posted to Panama on her first venture with the American Armed Forces. She was supposed to stay there for an entire two-year tour, but Cathy's restless spirit again changed her fate. She did not like Panama and applied for a position with Military Academy Preparatory School (MAPS) in the autumn of 1984. She was accepted and finished that program.

She and the other students now had to decide either to attend West Point or go on to a new duty station. Cathy chose the duty station. She was further given the choice of serving at the Presidio of San Francisco, or Fort Campbell, Kentucky. She chose Fort Campbell. Any other choice would have taken her away from the fateful morning of December 12, 1985, at Gander.

Cathy was attached to a medical unit at Fort Campbell, where she used her recently acquired trade skill to repair and service field generators and other equipment. In 1985 a contingent of the 101st was put together to serve a six-month tour in the Sinai Desert. The group would relieve the 10th Mountain Division of the Multinational Force and Observers. Cathy's medical unit was not to be included in that task force. Cathleen herself would not even be eligible, as she was due for discharge from the Army early in 1986. Again the woman's fierce determination decided her fate. She wanted to go to the Sinai. She applied for the one available position in her field and was accepted.

Virginia Ruth Word was Cathy's best friend. They had met in the military and soon become as close as sisters. The pair were soon given nicknames, "Jenny" and "Ziggy." When not on duty they toured the Middle East. Cathleen was a spiritual woman and not ashamed to show it. She was baptized in the Red Sea, to be closer to Christ. On the day of her baptism, Jenny presented Ziggy with a Bible and signed her name on the inside cover.

Virginia was twenty years old. She was also an honours student, and this beautiful brunette had dreams of becoming an actress. During her school years she played roles in all of her school plays. When Jenny graduated she wasn't ready for college. She had had enough of school for a while. Instead, she decided to join the Army for at least one tour. She did so against her parents' wishes.

The two women soldiers were pleased with what they had accomplished in such a short time. It wasn't every woman who went to the eastern deserts of the ancient world. They had worked alongside the men and had proven themselves. They had been accepted by the others and were now an integral part of the Screaming Eagles.

On the morning of December 12, 1985, Jenny and Ziggy strolled around the Gander terminal building. They bought a few small items in the gift shop. They listened to the age-old Christmas music piped through the ceiling. They were going home at Christmastime and their futures were bright and endless. It was a good time to be alive. When the boarding call came, both soldiers walked side by side toward the finger.

Following behind was the third and only other female in the 101st Airborne Division. Sergeant Christine M. McCleery walked quietly along with the others. Down the long corridor the three women went with the others, following the pointing finger to whatever lay beyond.

* * * * *

ROBERT STANLEY KAPLIN LOVED the Christmas music coming from the open steel rafters in the Gander terminal. He loved music, especially Christmas music. The familiar strains gave him a warm, peaceful feeling. Several of the soldiers were singing along with the tunes. Robert hummed along quietly, relishing the sweet euphoria of the returning soldier. They were almost home. Robert was from Tacoma,

Washington, which was known as the City of Destiny. The moniker was not lost on Robert when he joined the 101st.

He strolled around the spacious terminal with the rest of his battalion, happy to be on his feet after the long transatlantic flight. Everywhere he looked were small replicas of old planes, old artifacts, and memorabilia encased in glass displays. Photos of pioneering airmen, some from his own country. He had never seen such a display in any other airport. He knew about Gander's aviation history and was eager to know more. Robert read every plaque and studied each air machine on display. He kept hoping he would find a B-17 American bomber among them.

Robert knew just about everything there was to know about B-17s. The B-17E aircraft was American-built, by Boeing. Also known as the Flying Fortress, owing to its fierce fighting reputation, the bomber began its history on July 28, 1935. It soon became the pride of the American Air Force. It was also used extensively by the Royal Air Force of Great Britain. As late as 1968 the Brazilian Air Force was still using the intrepid B-17. Boeing had built no less than 12,731 of them. Its primary use was a deadly one: the B-17 was a war machine.

Robert's grandfather had flown in one during World War II. He was the navigator in one of the B-17s that were involved in the massive bombings of Nazi Germany known as the Berlin Raid. An incredible one and a half million metric tons of bombs had dropped on German soil through the course of that war, 640,000 of which were dropped by B-17 bombers. Robert's grandfather had navigated bombing runs and drops for twenty-one consecutive night raids.

Just a few hours ago, while waiting for the Arrow to be

refuelled in Cologne, Robert couldn't help but wonder about his grandfather's involvement in the destruction of Adolf Hitler's Germany. Near the eastern edge of that country's border with Poland, where all of the conflict had begun, Berlin, Germany's capital, had fared the worst. The Allies had pounded that city into the dirt during the closing days of the war. Berlin's destruction was complete. Robert's grandfather had been presented with the Bronze Star for his part in the attack on Berlin.

It came as no surprise to anyone when Robert had joined the Armed Forces. He had made that decision when he was only twelve years old. Robert intended to make the Army his career. His life. He would gladly follow in his family's footsteps. His father, Walter, had served his country the same way as did Robert's grandfather. Walt was an Army man. He had served with the 504th Airborne MP Unit out of Fort Gordon, Georgia.

Robert wondered if the war machines of today would one day be displayed as antiques in glass cabinets in this same Gander terminal building. He headed for the finger with the others, his destiny forever entwined with theirs.

Sergeant Robert Kaplin didn't know it, and indeed would never know it, but in time he would receive the same Bronze Star Medal as did his grandfather.

CHAPTER 6

I WENT BACK TO the Silent Witness Memorial again after I had written one half of the manuscript *Where Eagles Lie Fallen*. It was December 12, 2009, the twenty-fourth anniversary of the disaster. My wife and I were attending the annual ceremony held on the site. The flags fluttered at half-mast. The day was cold, -7°C, but sunny and very bright. The sky above and the lake below held the deep blue of the coming winter.

A small crowd had gathered. The Gander town mayor, Claude Elliot, was there. The Canadian Armed Forces had a small contingent there. Scott Simms, Liberal MP from Parliament, and Progressive Conservative MHA Kevin O'Brien stood among the dignitaries. The Salvation Army Gander Citadel Band were there. A contingent from 9 Wing Gander CFB was mustered. The Royal Canadian Legion attended. The Canadian RCMP, dressed in resplendent suits of red and topped with broad-rimmed Stetson hats, stood ramrod straight. The Gander International Authority, the Gander Fire Rescue, and even the North Atlantic Aviation Museum were represented. All of them laid wreaths. I saw no U.S. representation.

First the Canadian and then the American national anthems played. The dignitaries gave short speeches. The colourful wreaths were laid down, and there was a prayer. Bandmaster Calvin Way blew the *Last Post* and then *Taps*, which was made famous by the U.S. military. The tune is played regularly at flag-raisings and funerals. The bugler's lips showed white in the frigid mouthpiece as he blew the haunting tunes into the cold, windy air. A few of the lesser-known lyrics of the old tune *Taps* came into my head.

Thanks and praise, For our days,
'Neath the sun, 'Neath the stars,
'Neath the sky,
As we go, This we know,
God is nigh.

During the minute of silence that followed, the flag lanyards rattled softly against the steel spars. The sound of a truck on the highway faded away. And then it happened. A child cried. Jacob Cayer was crying for his mother. Sergeant Jeanie Hancock looked nervously around for her boy, who was in the arms of his father. Master Corporal Jon Cayer was standing apart from the rest of the CFB troop and was trying his best to quiet his whimpering son. It wasn't working.

The cry, to me, was penetrating. It wasn't an angry sound. It was a soft, pleading wail for his mother, who was standing just out of his reach. It was a compassionate human sound, unashamed and sincere. It was a "But why can't I go to my

Mom?" cry, which was soon answered by a mother who would not see her son suffer, Army or no Army. Jeanie stepped out of rank, walked to her husband's side, accepted her son, and walked off the site. I followed her, notebook in hand. For me it was the most moving part of the ceremony. A small child had unknowingly stirred within me what all the carefully prepared adult speeches and detailed planning could not do. It was the simple lament of one human being missing another. The sound of youth and continuation, in this place where death had come. The silent witness voiced.

* * * * *

CHRISTINE KLINEFELTER MANION HAD a terrible premonition that her husband, Captain Edward J. Manion, was not coming home. Edward was stationed in the Sinai with the 101st Airborne Division and was scheduled to come home on December 12, 1985. The uneasy feeling persisted in Christine's head. It was draining, an unexplained feeling of dread for her husband. It haunted her nights and persisted through the daytime. It was also a feeling she couldn't really share with anyone.

She decided not to go to a welcome-home ceremony for the returning soldiers. She hoped by doing these things her mind would stop sending her such crazy signals. Her husband had always come home. He would come home this time. The night of December 11 was the longest one of Christine's life.

Unable to sleep because of the all-consuming fear of her husband's death, she took sleeping pills in the small hours of that night. When she woke her man would be home and her troubled mind would be at rest, she knew. But when she finally awakened from her induced, prolonged sleep, her woman's premonition was plastered all over her blaring TV screen.

* * * * *

GLENN BLANDFORD LEANED BACK in the seat of his silver Honda Accord. He had parked as close as he could get to the north shore of Gander Lake. He pressed a button and the window rolled down. He wanted to hear the noise the water made against the shoreline. It was one of his favourite sounds.

The northeast wind blew coldly through the open window. *My God, will it ever warm up?* Glenn thought. Still, he reasoned, it was only early May; nothing unusual about cold, wet days for Newfoundland. He debated getting out of the car and decided he would, despite the raw wind. Pushing his 215 pounds out of the seat, all six feet and two inches of him stood against the wind. He zippered his light jacket close to his chin and wished he had worn his winter one as he stepped away from his car. The lake was high, though not as high as other springs. Already the winter melt had drained from the ridges. It was mid-May 2010.

Glenn was born in Gander in 1947, in the Banting Memorial Hospital on what was then called the "Army Side" of town. His parents, Sam and Nellie, had come to the new community that had been forged out of the wilderness, in search of a better way of life. The Blandford name was steeped in history farther south; a town near the coast of Newfoundland bore their family name. Port Blandford was a small outport nestled away at the very end of Clode Sound. The town, deep inside the fjord of water winding in from Chandlers Reach, was more than twenty miles from the open salt sea.

Glenn had visited the town of his ancestry many times. The only difference between the waters of Gander Lake and the bay at Port Blandford, he decided, was the salt, the landlocked shoreline of the small seaport resembling nothing more than an inland lake.

Glenn loved the sea, and not just the quiet bays and sheltered coves. He loved the wide-open sea with all of its danger and mystery. Its endlessness. He had sailed on the open Atlantic many times, around the rugged coast of Newfoundland and beyond. Up along the American seaboard he had sailed, a willing part of the crew that brought a fifty-foot yacht that was his father's pride. They had encountered dense fog and high winds. Glenn loved it all.

Here by the quiet shores of the huge Gander Lake, with a brisk wind rushing the clear breaking waves upon the rocky shore, Glenn was always reminded of the sea he so dearly loved, and of the father he missed so much. His father had been the chief mechanic as well as the vice-president of Allied Aviation. He was also a pilot and flew his own plane.

Glenn had been coming here to this spot all of his life. He had brought his family here for picnics, for hiking, for boating. But there was a time when he couldn't bring himself to come here anymore. For more than two years after the Arrow Air crash of December 12, 1985, Glenn had not driven down this road.

Midway down the ridge behind him, the road passed through the spot where all of those Americans had perished.

Glenn Blandford was the last man on the ground to speak to any of them before they died. It was a terrible record to hold. In the years since the crash, he had done many interviews concerning that fact.

Now here it was again. Someone else wanted him to tell his story. Would it never end? This year, 2010, was the twenty-fifth after the Arrow Air had crumbled here, close to the shores of this peaceful water.

Turning his back to the cold, Glenn climbed back into his car, backed away from the small parking lot, and drove slowly up over the hill. When he passed the Silent Witness Memorial, he wondered, as he had done many times before, why it was he had not been able to come here for two whole years after the disaster. He had a clear conscience about the whole thing. He was a trained, methodical, professional, and very practical person. He rarely let his emotions decide anything for him, but for all those months after the crash, he couldn't come down the quiet road to the shores of Gander Lake. He still didn't know why.

Glenn had a physical reminder of the Arrow Air crash in his backyard.

Workers had hastily erected some temporary structures on the crash site to accommodate their needs, as well as to house the equipment used in the cleanup. After the site work was deemed finished by the authorities, the structures were dismantled. The lumber used in the makeshift buildings was sold cheap and recycled. Glenn had purchased enough of it to build a shed in his yard. Every time he walked into the shed, he remembered the Arrow and his connection with its final moments. All these years since, his shed with its "Arrow lumber" was still there. So was his memory of that morning. He would tell his story one more time.

* * * * *

ON THAT COLD, FOGGY morning, Glenn Blandford was the sole air traffic controller in the tower at Gander International Airport. There was nothing unusual about his being alone at the tower. Air traffic at the airport wasn't as busy as it used to be. Blandford had started work at midnight, the graveyard shift. The tower room was filled with screens and electronic equipment. Soft green and red lights displayed to the knowing eye what was happening high up in the skies over the sleeping town. The ambient lighting was dim and unobtrusive. No bright lights to mar the vision of the controller who had the responsibility of guiding hundreds of lives to the safety of earth.

During the night it had snowed a bit and some freezing rain had fallen, but it hadn't amounted to much. The snow-clearing crews had quickly taken care of it. Blandford could see them working from his commanding view in the tower. The temperature outside had been hovering around the freezing point all night. There was little to no wind, but plenty of fog. Air traffic had been slow during the night. Only four planes had needed his attention and direction.

Glenn Blandford had worked at the Gander tower for fourteen years. After an intensive six-month training course in Ottawa, he had started work here on July 6, 1970. It wasn't a job he took lightly. Every landing and takeoff was given his utmost attention. On this night, and until the day shift took over, he had handled ground control along with his normal air control. It was all part of the job to him.

At 5:00 a.m. Blandford received the runway conditions. Runway 22 was close to sixty per cent covered with ice. By now the trucks were spraying the surface with an ice-cutting chemical. The ground crews at Gander International Airport were more experienced with ice and snow than anywhere else in the aviation world. The island of Newfoundland is anchored at the merger of two of the world's mightiest ocean currents. Cold Arctic waters brought south by the huge Labrador Current are capable of moving mountain-high masses of ice; they collide with the warm waters of the Gulf of Mexico – the world's largest gulf – just offshore, and the coming together of these great ocean-travelling streams creates atmospheric changes that are sudden, unpredictable, and unique to this island.

Blandford picked up an aircraft coming in from the east on his radar screen. *Right on time, as always,* he thought. A few minutes passed and then a voice spoke loud and clear in his headphones.

"Good morning, Gander. This is Big A 950 inbound your space from Cologne, Germany. Requesting landing clearance and instruction." The voice of Captain John Griffin sounded cheery. It was the DC-8 Arrow Air. They always referred to themselves as the "Big A." Blandford had been expecting the flight. Arrow Air was a regular visitor in Gander; the plane always stopped for housekeeping here.

"Big A 950, Gander Tower. Welcome back. Advise left turn to west. Cleared for landing on Runway 04. Acknowledge."

"Gander ground, Big A 950. Glad to be back."

Blandford followed the aircraft's progress on his screen as it turned into the new heading. Minutes later it roared to a landing and into his view, its cabin lights flashing against the black runway.

"Big A 950, taxi via Runway 13. Park Gate 8."

"Gander ground. Roger that."

Arrow Air pulled up to the gate, the door opened, and the soldiers stepped out into a raw Newfoundland December morning that was nothing at all like the Sinai Desert. It was 5:45 a.m.

Blandford pushed his chair away from the control desk and removed his headset. Another routine night and early-morning landing was done. For the next thirty-five minutes or so, Glenn looked out the black windows at regular intervals and watched the activity below the tower. Two fuel trucks filled the DC-8's tanks. He saw the Arrow Air's flight engineer step down from the plane, sign for the fuel, and walk carefully around the aircraft. The man used a bright flashlight to inspect the plane. Bags of garbage were removed. The waste tanks were pumped dry. *General housekeeping*, Blandford noted. The outside temperature still hovered around the freezing mark. The plane was not de-iced.

At 6:15 a.m. Blandford pulled his chair back to the control panel, sat down, and placed the headset on his head again. Below him the first of the soldiers were emerging from the finger. They walked across the tarmac two abreast. Even from here Blandford could see the mood they were in. He could tell by the way they walked that some of them were women. Some looked up at the tower, their faces appearing to be smiling. A few of them seemed to be mock dancing along. One soldier slapped the shoulder of his buddy. They both appeared to laugh at some remark. It was a joyous time for them. It was Christmas and they were going home.

One by one the members of the 101st "Band of Brothers" walked away from the terminal, climbed the portable steel ramp, and stepped inside the waiting airliner. For a second each one of them was highlighted in the glow from the cabin entrance. The man in the tower watched as, one by one, the peacekeepers from distant Sinai disappeared inside. The big door slammed shut and the ramp was pulled away. Dark shadows appeared before each small oval window along the length of the Arrow as the soldiers found their seats.

Blandford's headset burst into sound. He had been waiting for it.

"Gander Tower, Big A 950. Heading for Hopkinsville, Kentucky. Our last leg to home. Clearance on request, please."

"Big A 950, no clearance delay anticipated. Begin taxi when ready."

"Gander Tower, Big A ready to taxi."

"Big A 950, head straight ahead Delta to Runway 13. Taxi to Runway 22, turn right and stop. Winds are calm."

Captain Griffin confirmed his orders. "Runway 22 via Delta and Runway 13. Roger that."

Two blips appeared on Blandford's screens. Planes coming up from St. John's. The closest one was a Navajo twin-engine. It was still twelve miles out and requesting landing clearance for Runway 31, which was on the opposite end of Runway 13. The lights from Big A were clearly visible as it taxied along Runway 13.

Blandford spoke into his mouthpiece. "Big A 950, Gander Tower. Incoming traffic twelve miles out requesting landing on Runway 31. Can you expedite Runway 13 taxi?"

"Will that affect our takeoff, Tower?"

"We will get you away before the incoming gets down. I just want to get you safely clear."

"Roger that."

"Go ahead."

Blandford's headset crackled again. "Gander Tower. This is November Lima Tango 351."

Glenn Blandford spoke to the rapidly approaching aircraft. "NLT 351, Gander Tower. Runway 31 for you. The winds are calm."

"Thanks, Gander Tower."

He switched his focus back to Arrow Air. "Big A 950, Gander Tower. Make a right turn now on Runway 22 to takeoff position, when you are ready for takeoff."

"Gander Tower, Big A 950. We are ready for immediate departure."

"Big A 950, you are cleared for takeoff on 22. Winds are still calm. Merry Christmas."

"Merry Christmas, Gander."

Blandford's headset went silent. He looked out and saw Arrow Air racing along Runway 22. No matter how many times he saw it, the silhouette of a night plane racing along the edge of the earth seeking the freedom of flight still fascinated him. The Arrow's lights disappeared from view as the plane went down the slope near the end of the runway and reappeared again in full view as the craft lifted off and climbed into the grey morning sky.

Inside the steel belly of Arrow Air, the men and women were no longer soldiers and crew. They were all humans headed for home. It had been a long flight from the Middle East, across the face of Europe and over the broad reach of the North Atlantic.

There is a feeling about home and family that comes only after a prolonged absence from it. It is a time when the sweet sorrow of parting is not only forgotten, but it is worth it. Images of small, running children with open arms, and expectant wives and sweethearts shy among the waiting crowd, bring smiles and misty eyes to the pilgrim nearing the journey's end.

Aboard the Arrow the engine's whine increased. Seatbelts were fastened. Heads settled back against the high seats. A slight shiver ran through the fuselage. The motors thrummed, and the Arrow Air continued on its collision course with destiny.

* * * * *

GLENN BLANDFORD SAW THE Arrow leave the ground, the dimmed cabin lights trailing a smoky light along the length of the plane. Only a few seconds into the air and the plane was levelling off. Glenn stared out the window. Why wasn't Big A still climbing? He swivelled in his chair, his eyes zeroing in on the radar screen. The Arrow wasn't there. He whirled around. There were no lights from the leaving Arrow. Then a glaring, sickening burst of flame rose above the dark treeline. It was followed by an awful explosion. The Eagles were down. It was 6:55 a.m.

Instinctively he pushed the CRASH FIRE button. It would alert the Transport Canada team. At the same instant the phone rang. Transport Canada was calling him. They too had seen the flames and heard the explosion. The response to the Arrow's terrible beacon had begun.

Glenn's attention was pulled away from the glowing sky to the west. An Eastern Provincial Airlines flight coming up from the east needed his attention. The pilot had seen the explosion. Glenn asked him if he would make a turn to the west, to identify where the flames were coming from before landing on Runway 31. *The explosion could have come from the lake*, he thought. The EPA flight turned as requested and soon confirmed the flames were very bad and were coming from the forest just west of the Trans-Canada Highway.

Tom Scott, the tower boss, called Glenn. He was to be relieved of duty immediately. It was the protocol for Transport Canada.

Glenn stood by the tower window and stared down the length of Runway 22, where steaming breaks could be seen among the blaze as firefighters battled it. For one terrible second his active mind tried to create the scene of a plane down at the end of Runway 31. In the residential part of town. Where his family lay sleeping.

He banished the image from his head. He glanced once more at the panels and screens. The control one man could have over so many others came to mind. Before he left the room, it suddenly came to him that this time it wasn't enough. With a sudden, terrible urge to see his family, Glenn Blandford walked out of the Gander control tower and headed for home.

CHAPTER 7

It was three o'clock in the morning, and Sandra Kelly couldn't breathe. She was going to die, and she knew it. A thick, noxious smoke filled her throat. She coughed uncontrollably, trying to free her fevered lungs from the smoke that would kill her. The sounds that invaded her head were horrifying. Her eyes watered and burned. She could hardly see anything.

Screaming men were torn from high-backed seats and vanished, still screaming, from her blurred vision. The sounds of shattering glass from dozens of small oval windows added to the cacophony of Sandra's terror. Hundreds of terrified voices all blended into one dirge-like wail of death. Sandra was aboard Arrow Air as it came plunging down. And over it all was the din of out-of-control engines and rending steel. It sounded as if a thousand race cars were careening, unchecked, down a straightaway.

A black figure came from nowhere and landed across Sandra's legs. It looked like a black manikin with limbs twisted at grotesque, abnormal angles. The manikin moaned. A scream louder and higher-pitched than any of the others tore into her senses. It was a sound of pure terror emanating from the emotional depths of a woman who knew she was going to die a horrible death. It was Sandra's own scream that finally awakened her from her nightmare.

Her husband held her soaked, trembling body in the darkness of their bedroom. She was still screaming. Her cries had awakened her two boys. Five-year-old Sean was peeking around the waist of his nine-year-old brother, Ron. The sight of her frightened children brought the stricken woman back to reality. They had never seen their mother cry before. Outside, the cold Newfoundland winter pressed its blackness against the windows.

The Arrow Air disaster had taken place weeks ago. Christmas had come and gone. A new year with all of its celebration and expectation had begun. Sandra Kelly had not shed one tear the day the Arrow fell from the sky. Nor did she cry during any of the days that followed, even when she was among people who were openly weeping. Sandra wondered why she wasn't crying with the rest. She had even thought that she had suddenly become a hard-hearted woman. Her husband had put that thought to rest: he said she was the most caring and big-hearted women he knew.

Now, sitting on the edge of their bed with the arms of the man she loved holding her, Sandra wept uncontrollably. She wept for the men and women who had not reached home. She wept for the sudden loss. She wept for what she had seen. For the hopelessness. For the children who waited. She wept because she had frightened her children. She wept because she had not wept before.

The barrier to express her grief had been removed. Sandra feared the nightmare, but secretly she was glad she was weeping. It somehow cleansed her from the horror she had witnessed the day the Arrow had fallen from the sky on the edge of her town.

* * * * *

FOR SANDRA KELLY, DECEMBER 12, 1985, had started with two telephone calls. The first one was for her. It was 7:00 a.m. A plane was down, she was told. She was needed at the airport. Sandra was the deputy mayor of the town of Gander. The mayor, Doug Sheppard, and several of the town councillors had gone to St. John's on town business. Sandra would now play the unexpected role of mayor. It was only a month to the day since she had been elected. She wasn't sure what she would have to do or what was expected of her. She figured it was an Eastern Provincial Airlines plane. It probably had a full load of passengers.

The second call that morning, just a few minutes later, was for her husband. Ron Kelly was a doctor who worked at the Gander hospital. He too was told there had been a plane crash. He was instructed to return to the hospital immediately.

Ron and Sandra made hasty arrangements with a babysitter for their boys, then left to deal with their respective emergencies. Sandra was also a nurse, but this morning her services as a town official were needed. Before this day was over, Sandra would come face to face with more death than any she had ever seen in her profession.

Sandra Kelly was an attractive woman with a full head of

brown hair, always neatly groomed. She walked with a firm, quick, confident step. When she arrived at the Gander terminal, she suddenly remembered she hadn't combed her hair. Her step slowed.

The Emergency Control Centre was buzzing with people when she arrived. As town official, Sandra expected to be brought immediately into the centre and be briefed. She wasn't called or even allowed inside the ECC at first. Even later that day, when the media arrived from all over the world along with American officials, a high-ranking American military man singled her out in the crowd, pointed his finger at her, and said in a loud voice, "I hope you take good notes, young woman." Sandra responded in her bravest voice. "I am not a secretary, sir. I am the acting mayor for the town of Gander."

The man never acknowledged his error.

* * * * *

INSIDE THE ECC THE magnitude of the crash was revealed. The flight was confirmed as Arrow Air. There were 256 people aboard, 248 of whom were American soldiers. The other eight were plane crew.

Out at the crash site, the firemen were battling the raging fire that had consumed Arrow Air. The reports kept coming in to the Emergency Control Centre. There were no reported survivors. Inside the ECC itself every possible emergency had to be addressed. How close was the aircraft to the lake? What else did Arrow Air carry besides passengers? Were there chemicals or toxins aboard? Would the lake be contaminated? Gander Lake was huge, but the downed plane was not far from the town's water intake. Someone located a boom and it was taken around the shoreline directly below the crash site. Ordnance experts were contacted. The town of Gander was probably the most prepared airport town in North America. The only things the town was short of this day were body bags and morgues.

Pat Kane stood as close to the wreckage of Arrow Air as he could manage. He knew that everyone aboard was dead and that most of them were men and women from the American Armed Forces. Pat was technically an American himself. He was born in Hurleyville, New York, where his father – born in Gambo, Newfoundland – and his mother had worked for a time.

Pat had walked close enough to see just one body. It was a young man with his back slumped against a small tree. He looked unharmed, but he never moved. His face was a deathly, unnatural white. Pat turned away and walked back toward his car, suddenly realizing that he was almost creeping along, as if not wanting to make a sound. The place was unnaturally quiet. Men who moved around in the area talked in whispers.

Pat Kane was a big man, six feet tall and a little over 200 pounds. Everyone who knew him always remembered his constant smile. Pat wasn't smiling today. He also felt very small and unprepared for so much unexpected death. Hired by Transport Canada to taxi officials and others to the site, Pat had had a busy day. He had just brought Sandra Kelly to the site, but he had no intention of following the deputy mayor down into "The Hole."

* * * * *

DR. PETER MATTHEWS WAS twenty-eight years old and only three years out of medical school. He and his wife,

Patricia, had been married for just one year. Matthews enjoyed working at the James Paton Memorial Hospital in Gander and looked forward to many years of medical practice. He was already a part of the hospital's DTT, or Disaster Triage Team.

The small town of Gander had always been prepared for any plane crash. Matthews had never been involved in a major disaster before. The phone call he received at 7:00 a.m. on December 12, 1985, alerting him of a large airliner down somewhere east of town, was disconcerting to the young doctor. He got ready with the rest of his team at the hospital and loaded their medical gear into a Transport Canada vehicle before leaving. They were told that, according to all reports coming to the hospital, there would probably be no survivors.

Just over a mile to the east along the Trans-Canada Highway, Matthews and his team prepared to set up an emergency triage. It took some time before they were allowed closer to the site; the area had to be secured for safety reasons. The fires were still burning, and further explosions were feared. The officials on site told the doctors they had found no survivors.

The four physicians and as many nurses that made up the triage team had been prepared to administer the first lifesaving measures to badly wounded people, but there were none. No stretcher-borne soldiers in need of their help, no cries from victims trapped inside the wreckage that was once Arrow Air. Their medical bags, filled with everything needed to evaluate and treat all kinds of injuries in the field, were not needed. The DTT decided to walk the entire length of the Arrow's destructive path in case there was life hidden somewhere. They were told to document any and all human remains. They were also told to inspect any bodies they found for signs of life. They were not to move or disturb any human remains.

Peter and the others walked carefully along the debris trail from the Arrow. Everything was quiet. He had mentally prepared himself for shock, but the whole scene was more surreal to him than anything. He realized that if there had been multiple injuries and the cries of human suffering that went along with them, his reaction might have been entirely different. Matthews saw many dead bodies as he followed along that terrible route. When he and the others reached the place where the Arrow had finally come to rest, they knew their services would not be needed. Not one stethoscope was pulled from one bag. Doctor Peter Matthews, trained to save lives, felt frustrated and even cheated. He and the others of the DTT simply walked back up the hill. Their services were not required.

* * * * *

RON DELANEY LIVED ON 1 Wood Crescent in Gander. It was a quiet, residential street just a short drive west of the airport. Ron was an electrical draftsman by trade, and he was also the fire chief on the town's volunteer fire brigade. He was born in the small village of Northern Bay on February 22, 1947. Northern Bay is situated on the north tip of Conception Bay, on the small Bay de Verde Peninsula of Newfoundland's huge Avalon Peninsula. Ron's family had moved to Gander when he was only nine months old.

Ron and his wife, Linda, were married with three children. They seldom slept late. It was just past 6:30 a.m. on Thursday,

December 12, 1985, a school day, when the phone rang in the kitchen. It was Chester Henniker, the fire chief with Transport Canada. Henniker was Delaney's boss in the event of a crash near the airport, and if the volunteer brigade was called. Chester didn't waste time with pleasantries this morning.

"Ron, there has been an explosion, a big one. A plane is down, Arrow Air. It appears to be just off the west end of 22, this side of the lake. Hurry." He hung up the phone.

Ron was also the designated crash fire chief with Transport Canada. He rang the one emergency number that would be relayed simultaneously to all of the town's volunteer firemen's pager systems, and rushed out the door.

Driving his red Chevy station wagon, with the Town of Gander Fire Brigade crests on both front doors, complete with the bold words FIRE CHIEF underneath the crests, Delaney headed for the downed plane. He didn't use the streets leading to the highway, just beyond which a huge fire lit up the north sky. He drove straight down Runway 22, speeding over the same stretch of concrete Arrow Air had just recently soared across.

Ron wondered what he would have to face this time. This was not the first time he had witnessed or been directly involved in a plane crash. There had been a few small ones, but the one still fresh in his mind as he headed down the runway was the Czech crash on Tuesday, September 5, of 1967. Delaney would never forget it.

The Russian-built Ilyushin 18d turbo-prop airliner had left Prague in Czechoslovakia for Havana, Cuba, and refuelled at Shannon, Ireland, before crossing the Atlantic to Gander. Her tired crew was changed at Gander, after which she refuelled and proceeded to leave for Cuba. When she lifted off of the 8,900 feet of assigned runway, air traffic control personnel thought something was wrong. To the experienced people in the control tower, who had watched thousands of flights land and leave, the Czech plane's liftoff looked too shallow. They were right. Too late, the plane tried to veer away from a radar pole – her right wing tip made contact with it and exploded. Before the unbelieving eyes of the people in the tower, the Czech plane swung like an acrobat who had missed his thrown line. The left wing, as well as the engine props on the same side, tore a swath through the Canadian National Railway steel tracks and its crushed-stone bed, and the plane somersaulted into the soft bogland of the Gander plateau, where it burst into flames. Thirty-two on board died instantly. The first firefighters on the scene were shocked by what they saw. Ron Delaney was one of them.

Thanks to the joint, rapid response of the Gander volunteer firemen, the airport fire crews, and both the Canadian and American Forces then stationed in Gander, thirty-two of the passengers were pulled alive from the burned wreck. The James Paton Memorial Hospital in Gander, as well as the Central Newfoundland Hospital at Grand Falls emergency rooms, worked ceaselessly to save the badly injured survivors. Despite their best efforts, five more people would later succumb to injuries sustained in the deadly crash.

Forcing that long-ago image from his mind, Delaney neared the west end of the runway. He slowed just a little and pulled off the paved surface, his speeding car swaying as he went down the slope through the opened gate in the steel fence surrounding the airport. The gravel strip he drove upon led to and across the Trans-Canada Highway. Two tractor-trailers were stopped there, one in the eastbound lane, the other in the west. Delaney plunged across the highway and down the gravel road toward the sky-lined inferno. This was not going to be pretty.

His two-way radio squawked. His crew were leaving the fire hall. Ron spoke into the mike. He ordered a group of five and one truck to remain in town; he had to leave a backup behind, even though he knew this was going to be a bad crash. The town still had to have fire protection. It was always the rule.

Delaney could smell the burning jet fuel before he opened the car door. Dense, black smoke billowed upwards from a raging inferno, the flames tearing through the smoke in long, jagged fingers of death. There was another smell more foul than that of the burning fuel. Ron knew what it was. When a man got the stench of burning human flesh into his head, it never left him.

Aside from the licking flames, nothing else moved. A few trees were burning, but strangely enough they made little sound. Nothing moved. There was no life here. A wail of sirens filled the air behind him. His volunteers were already here. Ron glanced at his watch: less than eighteen minutes since he had called them. He suddenly knew that even if his pumpers were parked here at the moment of Arrow Air's violent impact, it would not have changed the outcome. This was to be a recovery without rescue.

Upslope from the burning wreckage of Arrow Air, outlined in cruel detail by the flames and the coming daylight, was a mowed and torn path as wide as the spread wing tips of the once proud aircraft. The tops of some huge trees had been sheared off neatly where the airliner had first made contact with the forest. Speeding to her death, the Arrow had forged a channel through the dense stand of trees. Some of the black spruce trees had been torn from the land, complete with their massive roots, and sucked along behind the doomed plane like weeds pulled from a garden.

Smouldering pieces of wreckage were everywhere, but the spot where the front of Arrow Air had finally ended its last landing, and had ploughed into this innocent land held Ron Delaney in place. Here, all of the plane's structure that had survived the plunge down through the tall timber had finally come to a full, crumbled stop. Everything that had stayed with the aircraft this far had been thrown with a whiplash of energy to the front.

The fire trucks arrived and volunteers went to work. They shut down their sirens. Transport Canada officials and RCMP machines and members of the nearby Canadian Armed Forces stationed in Gander all arrived with sirens wailing. They all shut their horns down as soon as they arrived. Men talked quietly as they applied water and foam to the burning pile that was now the focus of everyone's attention. Grey clouds of steam rose quickly, flooding the area with a sickly sweet stench that stuck to a man's throat and irritated nostrils and burned into his senses and forever infiltrated his mind.

The men started concentrating on several smaller fires. The truck drivers had to go to the lake nearby to replenish their water supply. A relay pumping system was quickly set up.

The pile of torn and twisted metal, rooted soil and rocks, and dead humans ignited again. Like a mound of cooling lava that flares up from within and leaks orange flame out through its deceiving black surface, the fire defied them all. The men readjusted their nozzles and applied more water. This time the stream of water continued long after the fire went down again. It would not ignite again.

The Gander firemen dragged their hoses all around the site, extinguishing fires large and small as they went. Under strict orders from their chief, they touched nothing. Everything that looked human received a ground peg and was noted. The scene was like a battlefield, the firemen like soldiers in the growing, murky daylight, moving soundless among their fallen comrades for any sign of life but finding not one.

Some of the victims were found away from the bulk of the wreck itself. They had been thrown from the descending aircraft. While it kept some of their bodies relatively intact, it did not save their lives.

One young soldier appeared to be resting with his back against a tree trunk. A fireman rushed toward him. The soldier did not move, and his wide-open blue eyes did not see the hopeful volunteer.

Delaney and his volunteers, along with the TC fire crew under Henniker, continued to battle the fire. The smell of raw fuel, and burning fuel mixed with burned and still-burning plastics, smouldering fabrics and melting metals with green and blue flames, acrid paint and black, smoking rubber made everyone nauseous, but dominating it was that other smell, one which none of them would ever forget.

A sweating, smoke-blackened fireman spoke to Delaney in a low voice. "I don't know if I can go back, Chief. I can't handle the smell. Its not just the actual smell itself. It's knowing what it is that bothers me."

"You do whatever it is you have to do. No one will fault you for not going back. You – all of you – have done the best of work under tremendous conditions. No one could do better. I have been to crash sites before, but nothing like this. You have no reason to be ashamed."

Delaney turned away and answered his radio. When he turned back the man had gone. Ron thought he had retreated back up the road away from the devastated area. Looking toward his crew, he saw the man pick up a hose and drag it toward what was left of the smouldering aircraft.

Several times during the terrible hours of that morning, others came to their chief and expressed similar doubts. All of them would forever remember this day as the worst day of their lives. A few of them would have nightmares for years after. But not one of the Gander volunteer firefighters shirked the task at hand, which was surely beyond their call of duty, on December 12, 1985.

* * * * *

MAURICE GEANGE WAS A paid firefighter with Transport Canada. He was forty-four years old and probably in the best shape of his life, but he still wasn't prepared for the carnage displayed before his eyes. He was the first man to walk toward the smouldering mounds that he knew were bodies. All of his training, which included hundreds of videos in living colour, and countless lessons from people who had experienced such crashes as this, didn't even come close to the horror he was witnessing.

He had watched the Big A roar down 22 and had seen the burst of flames light up the treeline a split-second before the explosion. The Transport Canada fire crash team directed by

Chester Henniker had sped down the same runway as the Arrow. The six-man fire team crossed the TCH and were now taking in full view what would haunt the rest of their lives. Incredibly the Arrow Air had been down for no more than five minutes. Still, Geange could see they were too late. They had all expected to see a burning plane, but what stared back at them was a searing hulk of metal.

Nothing could have saved them. The filled fuel tanks had ruptured and ignited. Driving down the gravel road that paralleled the destructive path of the aircraft, they could feel the terrific heat in the fire truck. Black smoke billowed skyward. The heat was oppressive, but the smell was the worst of all.

The fire crackled as the men applied foam from their truck. There was no other sound. No anticipated cries of anguish, no pleadings for help. There was not one moan of pain from even one of the 256 humans who had so happily boarded Arrow Air just a few minutes ago. Geange would always remember the scene that morning as one of an "alarming quietness."

On this morning Maurice Geange wished he wasn't a fireman. He and the rest of the firemen were walking among the dead. Maurice realized they were all talking in low voices. Only a scattered shout came from the men who were trying their best to knock down the fierce fire. They walked among the victims but moved or touched nothing. Maurice and Eddie Humphries walked side by side. They were searching and hoping for survivors. They saw some gifts, a few still in their Christmas wraps. Clothing and duffle bags. Books and personal belongings. Torn and broken bodies – smoking mounds curled up in fetal positions.

The firemen approached a group of bodies that looked like they could just get up and walk away. But there was no movement, no sign of life among them. Geange approached one young man who was sitting with his back to a tree. He was a black man. He looked unhurt. Geange hurried to feel the man's pulse. His fingers moved over the man's warm neck, seeking the rhythm of a carotid artery. There was none. Then Geange noticed the thin trickle of blood running out of the young man's nose and ears. He removed his hand from the soldier's neck, got to his feet, swallowed back his human emotions, and turned away.

* * * * *

THE FIRST THING DEPUTY Mayor Sandra Kelly saw as she neared the wreckage that had once been Arrow Air was a bottle of Johnson's Baby Powder. Her hand went to her mouth of its own will. As far as she knew, there were no babies aboard. The bottle was clean and untouched by the carnage that had taken place here. Just a short distance away from the baby powder was a navy blue shoulder bag. A carry-on bag, she guessed. It was also clean and intact. She wondered if the bottle belonged inside the shoulder bag. She had the sudden impulse to pick up the bottle and place it inside the bag, but she resisted and moved on. She walked slowly, not sure if she wanted to go any farther. Sandra was directed toward what the workers were calling "The Hole."

When the Arrow had come tearing down the forested ridge, it had plunged ever nearer to the earth as it came, uprooting and breaking everything in its path. When the aircraft made contact with the land below the treeline, it had pushed tons of debris before it. In effect, it had carved out its own grave in doing so. The whiplash effect from the final, sudden stop had brought the superstructure, as well as all of the contents of the airliner, into the hole created by the Arrow's bulldozing run. "The Hole" was far from empty. It was where most of the victims were found.

Sandra couldn't believe what she was seeing. As a nurse she had been expecting multiple injuries, burns, broken limbs, and severe trauma, but there was nothing here but death. *How many more seconds would it have taken for the Arrow to clear the earth and be safely away to the skies?* The question kept running through her head.

She noticed the firemen. Through the mist they looked ethereal, like murky actors who had gone quiet because they had forgotten their lines, as they moved silently among the dead. A few of the men had removed their hats and were carrying them in the crook of their arm. Seasoned firefighters already showing their respect for fellow fighters.

The image of so many still, twisted bodies reminded Sandra of manikins. Most of them were blackened by the fire that had raged through the downed aircraft. All of them lay as they had died, rigor mortis preserving their pitiful expressions. Back from "The Hole" she walked. Her quick step had returned. She wanted to put the scene behind her as quickly as possible.

She slowed again as she came back to the shoulder bag and the baby powder. Again she resisted the impulse to put the powder into the bag. She walked past without touching a thing. It was the Christmas month, she thought. A time for family sharing and celebration. It wasn't supposed to be about death.

This was the season celebrated for a special birth. It kept going through her head that Christmas bells would not be ringing for many American families this year. For them the bells would be tolling.

Sandra never shed a tear on that day. From the Arrow Air crash site she took away a scene and an odour that she would never forget. It was a clinging death-smell that took a hundred showers to cleanse from her skin, and one that countless waking mornings could not erase from her mind. It would take a dozen violent nightmares before the woman's pent-up emotions were finally released in a torrent of belated, pitiful tears.

CHAPTER 8

SERGEANT FRANK WARD WAS on duty the morning of December 12, 1985. He worked in security at the Gander International Airport. He was also Canada's representative to the American troops. Many considered Frank Ward as Gander's ambassador. Frank was an RCMP officer, a big man standing at six feet, four inches tall, all of it lean, tough muscle and bone. Frank had a heart to match his size.

Inside the Gander terminal on that dreary December morning, Frank was the epitome of Newfoundland hospitality. He wore the regulation working uniform of the Mounties. He shook hands with many of the young American soldiers as he walked among them, his smiling face beaming Christmas cheer. Sergeant Ward watched as the last man disappeared inside the finger that pointed them to the waiting Arrow. A few of them waved back at the tall Canadian Mountie.

Forty-five minutes later a different Frank Ward walked among the victims of Arrow Air. The smile was gone from his face, and his big frame seemed bent, reluctant to move any farther into this field of carnage.

The smiling faces of youth he had shared time with a short while ago were gone. In their place was nothing but a burning wreck of man's failed attempt at air mastery from which no human sound came. Frank Ward wished to God he were somewhere else.

* * * * *

WORKERS WHO ARRIVED FIRST had steeled themselves for what was to come. They had been forewarned. Horrible sights and smells and sounds that may haunt them forever. There were no sounds at all, just a deathly, unearthly silence that belied the terrible scene. It was as if nature and all of her wiles had chosen to silence the very earth upon which the dead lay waiting.

There were other scenes that tore into the senses and nauseated the stomach and welled salty tears into the eyes of hardened men and weakened them as sorrowing children. A young soldier with his hand – or was it a woman's hand? No one could really tell – wrapped firmly around the smoking trunk of a small tree. He or she must have somehow survived the initial crash of the Arrow and died in the fire that ensued.

* * * * *

DANGLING HIGH ABOVE THE dark green branches of one of the black spruce trees on the edge of the torn swath, a pair of dog tags awaited discovery. A young man below the silver amulet walked slowly along the devastated forest floor. He headed away from the tree, his head down, his eyes searching the uneven ground. He was looking for body parts.

He had been told to mark and report anything human. The young man hated his job but knew how important it was. He feared finding what he so carefully sought. All around him, people were doing just as he was doing.

Having reached the end of his search grid, he stood erect and stretched his aching neck. To his right several more men were searching the area just as carefully. He was now on the very edge of the treeline. He would have to look more carefully on his return; the destroyed brush and trees near the forced clearing was a good hiding place for the Arrow's secrets.

A faint glint high up in one of the trees back the way he had come caught his eye. Curious, he hastened toward the shining object. When he got below it, he found that the article of his attention was far too high to reach, and the lower branches of the tree did not encourage climbing. He started shaking the tree and was surprised when a dull, silver-coloured necklace tumbled down through the green-needled limbs and landed near the shadowed trunk of the tree. He knew it for what it was instantly. Picking up the small chain with its two rectangular metal pieces of identification, he was amazed to see the chain was unbroken. How could it be? He mouthed the engraved numbers aloud. When he finished he realized he was whispering. A cold shiver ran up his spine. It didn't chill him at all. He was sweating.

War had spawned the use of dog tags, and, oddly enough, a war fought in the Screaming Eagles of the 101st Airborne Division's own country. It was a war against itself. The civil war of 1861 in the not-long-united States, a deadly conflict waged entirely on its own soil. It was a war whose beginnings were so complex that historians today still debate its true cause.

It was on these battlefields that American soldiers pinned pieces of cloth and bloodstained paper to their jackets. Scrawled and sometimes even sewn on cloth patches were their names and those of their next of kin. Still other soldiers engraved their names on their belt buckles. Even in death, all soldiers want to go home. Before this home battle had ended, the U.S. government began issuing ID tags to its soldiers. The Franco-Prussian war of 1870 gave the tags a name that would stick forever. The city of Berlin ordered "marks on dogs" for its roaming canines. The tags resembled the ones issued to soldiers.

Sadly, here in Gander in 1985, the simple tags invented by their own countrymen could not identify all of the downed Eagles.

* * * * *

ONE OF THE LAST bodies was found underneath the roots of a tree.

There was only one possible explanation for the location of the body, a young man found with barely a scratch on him. The aircraft had uprooted the tree on impact as its great weight had slammed into the forest. In that same instant the soldier had been thrown from the doomed plane and was deposited under the tree's exposed moor. The softwood tree, still unbroken, sprang upright again seconds after the plane had raced past, entombing the soldier under its gnarled roots.

One of the searchers had seen a booted foot showing underneath the bushes surrounding the bole of the tree. He reached for it and was astonished to find it was still attached to a body. Surprisingly, the tree had required only a firm push by a few of the men to lean over and reveal the last pitiful secret of Arrow Air.

* * * * *

A SOLDIER'S BOOT, STILL securely laced, stood upright on the scoured ground. There was no soldier near it. A webbed belt with its Army-issue knife still attached was found in a cluster of low bushes.

* * * * *

A PLUMP BROWN TEDDY bear, its face singed, with streaks of oil tearing out of its black eyes, was found along the fringes of the site.

* * * * *

TWO SMALL DRESSES SPATTERED with black mud were found and handled by a teary-eyed volunteer, as gently as if the garments had been found covering the body of their intended child recipient.

* * * * *

A TORN OLIVE-SKINNED DOLL purchased in the souks of Egypt, its black eyes staring, was found splayed and tangled in one of the bushes, its gaily trimmed wrapping never to be seen by the small girl for which this Christmas gift was intended. A worker picked up the small doll as gently as if it were alive. He suddenly wondered if the child would even remember her father's face. He placed the sodden, ruined reminder of young life back onto the ground and turned his back away from the flames, away from the smell, away from the carnage.

* * * * *

A VOLUNTEER FIREMAN FOUND a folded piece of paper. Its edges were singed black and badly soaked from the firefighters' hoses. It revealed to the curious fireman the charred remains of a drawing. It could have been a charcoal drawing – the man wasn't sure. Nor could he be sure of the image the artist had so skilfully created. To him it looked like a magnificent hawk.

* * * * *

WOMEN STOPPED BY THE roadside offering food to the workers and even the use of their homes. For days after the crash, when the Trans-Canada Highway was open to free-flowing traffic again, motorists slowed as they passed the site. It was more a sign of reverence than idle curiosity. Women reached over and squeezed the hands of their loving husbands as they passed.

Volunteer firemen mustered in the Gander fire hall and talked quietly long after their work was done. None of them talked much about what they had seen.

The people of Newfoundland could be seen weeping as the news unfolded. One man, parked on the shoulder of the Trans-

Canada, was seen kneeling down in an attitude of prayer, while he wept unashamedly.

* * * * *

LONG AFTER THE GANDER crash, Michael Shayne Stack was finally identified by his dental records. His remains were returned to his grieving family in a small urn bearing his name and the dates of his birth and death.

* * * * *

TRACEY AND JEFF KEE were to be married at Fort Campbell on December 16, 1985 . . . which turned out to be the day of the memorial service.

CHAPTER 9

WILLIAM JEFFERSON BLYTHE III was born in the small city of Hope, Hampstead County, Arkansas, on August 19, 1946. His father had been killed in an auto accident three months before William was born. The boy would grow to be a man taller than six feet. The world would know him better as Bill Clinton. He would become the third-youngest, first Baby Boomer, and forty-second president of the United States on January 20, 1993. He was the city of Hope's most famous resident.

Mary Froelich was born in the same city of Hope, Arkansas, on November 3, 1935. She would never become famous. As an adult, Mary was only four feet, eleven inches tall. When Mary told me her height, I mentioned that she wasn't a very tall person. She laughed about it, then said something I would not soon forget. "I guess some of my troubles have pounded me down some." Mary had said it right.

She had six children. Only two would live to see their mother's hair turn grey. Two of her sons would give up their lives for their country. Nicholas Wheeler went willingly to fight for his country in Vietnam, that unwanted and most unpopular American war. He died there on September 13, 1970. Mary buried her first soldier son on October 12 of that same year. He was eighteen years old.

Mary's sister Nellie Johnston succumbed to cancer at the age of fifty-four on January 16, 1985. It would only be the beginning of Mary's troubles in that year of sorrow.

Her grandson Aaron, son of her daughter Dianne McKnight, was killed in a motorcycle accident on April 5, 1985. He was just eighteen years old.

Barely a month had passed before Mary and her family were visited with death again. Her twenty-two-year-old son Randy died of an accidental shooting on May 7, 1985. The small woman had indeed been given a big burden to bear.

Mary's youngest son, Ronnie, worked at a wheat silo. One day he climbed up the elevator shaft to replace a broken light bulb. He dropped the new bulb down the long shaft before he could replace it. Rather than use the stairs to go down for another bulb, Ronnie swung out on the escape rope. In the dim light he mistook an electrical cord for the rope. With great bravado he jumped for what he thought was the rope. The wire broke under his weight and he fell 185 feet to his death. It was June 19, 1985. Ronnie was sixteen years old.

Mary's other soldier son, Frank, came to his brother's funeral. Mary clung to her last son as if she would never see him again. Frank left after the funeral and was soon posted to Germany.

Mary Froelich was afraid for the time to pass, afraid what each new month would bring. Surely the tragedy she and her family had endured was enough. Her strong spiritual strength had nearly run out.

Her son Frank called from the American base in Germany and told his mother he was being shipped out to the Sinai with his battalion. He didn't know for how long, maybe six months or even a year. Time didn't matter to him. Frankie – his mother always called him Frankie – loved the Army. On October 3, 1985, Frank called his mother from the American base in the Sinai to wish her a happy birthday, even though he knew her birthday was a month away, on November 3. Frank told her he was being posted to another camp deep in the desert. He didn't know when he would be able to call her again. Rather than miss her birthday, he had decided to call now. Mary was very touched by the special attention from her only son. Before he said goodbye to his mother on that October day, Frankie made a strange request of her.

"If anything happens to me, Mom, I –"

But Mary interrupted her son. She had had to deal with too much death recently. "Don't talk of dying, Frankie. Children shouldn't die before their parents. I couldn't bear to lose you too."

"Oh, I'll be all right, Mom. I just have to tell you what I want just in case something happens – you know, years from now." Frankie convinced his still grieving mother to listen to his request. "Here's what I want done, Mom. On my tombstone, this is what I want written. 'Gone to Paradise to Be with My Brothers.'"

Mary listened without interrupting her son. She remained quiet after he had finished.

"Did you hear me, Mom?"

Mary's voice came from somewhere deep inside her. To Frankie she sounded as if she was resigning her future to fate. "I heard you, my son. And if I am alive it shall be done. I only pray to God that I am not."

Frank Wheeler's wallet and camera were found among the wreckage of Arrow Air almost a year after the crash. The Newfoundland winter and summer had barely tarnished these final remains. They were returned to his mother. Inside the badly burned wallet was a Roman road map. The camera held unexposed film. When it was developed the only distinguishable piece of the film held the image of the tomb of Jesus Christ. Frankie had taken the picture while in Jerusalem. Frankie was a God-fearing man who always read his Bible. One of the highlights of his life was walking through the Holy Land. His favourite place was the area where soldiers had laid to rest the son of that other Mary.

When Mary got the news on December 12, 1985, that her last son had been killed in a horrible plane crash in Gander, Newfoundland, she didn't scream or swoon. Mary simply slumped in her chair, as if in defeat, and began crooning gently. From that day on, she couldn't bear to be away from her two daughters, Dianne and Barbra Jean. She wondered if she would lose them too.

Entering the cemetery beside her husband and with her two daughters close at hand, Mary walked tall behind the flag-draped coffin. It was January 22, 1986. She buried her last son without even seeing his face. It was Frankie's twenty-second birthday.

* * * * *

BILL AND DOTTIE ZIEGLER buried their daughter, Cathleen Marie "Ziggy" Ziegler, on March 1, 1986. Her body was identified by process of elimination. The long wait for their daughter's remains brought little comfort to the grieving parents. The last words from their daughter had also been delayed. "Don't worry about me," she had written in her last letter home. To add to their misery, her parents had received that last precious letter three days after Cathy was killed.

CHAPTER 10

Twenty-two-year-old Bryan Gibson stared out the oval port window of the Boeing 737 at the rugged land below. From up here it was as if the big airliner was barely moving, the wide-angled view of the landscape framed with distance. They had come up over the southern edge of the blue ocean to find the rugged island blazing with the morning sun. Brian had never been here before. It was the summer of 2002.

He couldn't believe the size of the land below. He knew it was an island, but his idea of an island included low, hazy, palm-shaded atolls, similar to the ones made famous in hundreds of cartoons. Bryan was from the Great Plains area of America. He had never seen the ocean before. He knew, of course, there were big islands in the world, but it hadn't occurred to him how big the island of Newfoundland could be.

They had been flying over the land for more than an hour. Great flat-topped mountains rising out of blue fjords faded away behind the plane. Lakes mirrored the sky and carried the aircraft's shadow along. Countless marshlands and bare outcrops sprang out of dense forests. Rivers twisted through every visible valley, always headed toward the sea. No multicoloured patchworks of lush farmland here. No endless, identical blocks of houses fitted between straight, parallel streets. He was fascinated with the land's harshness, its diversity, its mystery, its beauty.

Bryan hadn't slept all night, and now, though he needed rest, his veins pumped with a strange mix of excitement and dread that brought him fully awake to stare eagerly out the window. The very name of Gander coming from the crisp voice of the captain stirred in him an emotion that he thought he'd been fully prepared for, long before this day.

"We'll be on the ground in twenty minutes, folks. A lovely seventy-one degrees in Gander," the captain finished with a faint crackle of static. The name Gander, Newfoundland, had been further branded in Bryan's mind just last year. The attacks on his country by the planes-turned-missiles was still on everyone's lips. Bryan had watched the TV coverage like everyone else. When he heard that a place called Gander was one of the airfields for craft seeking safe haven, Bryan had found out all he could about the assistance the residents of this small town had given his countrymen.

Bryan squirmed in his narrow window seat, his lap belt holding him back. Straining to better see the place that had consumed his life, he felt the airplane drop a little, settle, level again, and then bank to starboard, stealing his view of the land as the pilot altered course for the final approach. He tried to see out the opposite window across the tilted aisle. Other leaning heads blocked his view. The airliner tilted again, forward this time, angled down, and settled back in landing position. A great roar filled his head as the four engines were held in check and the ailerons whined as the crew put on the brakes.

A great green forest, amid open boglands and dozens of glittering ponds and meandering brooks, appeared below and beyond his window as the airliner steadied itself to meet the rising land. And then Bryan saw the big lake winding away to his left.

It was this image that had been etched in Bryan's mind: an aerial photo of Gander Lake. The water below glinted and flashed as the airship flew on, crossing the narrow body of water. He had stared at the newspaper layout for a long time in his room at home until he could identify the black-and-white image anywhere on the face of the earth. Now it was below him in living, intense colour. He shuddered with great emotion.

A long, wide clearing appeared just beyond the north side of the lake. The plane seemed to stall in mid-air as it approached the site, allowing him to absorb all of it. He thought he saw a flag waving in the clearing below. He didn't want to look any more, but he couldn't help but stare down at the fast-rising land where his father had died.

* * * * *

PAIN HAD COME EARLY to the life of Bryan Gibson. Barely five years old and with two spaces between his top baby teeth showing, he peeked up over the dark wooden sill of their street-facing window. A man dressed in the same uniform as the one in his father's framed picture on his mom's bedroom dresser was walking slowly up their driveway. It was not his father.

A long, black sedan was parked at the curb. It bore the Stars and Stripes emblem on its front door. Inside the car another man rolled down his window and waved to him in a slow, unsure manner. He was wearing a uniform too. The doorbell rang. His mother's step sounded light in the short hallway, her slippered feet producing a soft tapping on the polished hardwood floor.

Bryan heard the low murmur of voices from the door. He waved back at the man in the waiting car, who quickly looked away when he saw the boy wave. The man suddenly seemed to be trying to cover his bent head with his left hand. He rolled the window back up with his right.

A low-pitched moan came from the front hallway, followed by a wail of human misery and despair. It would haunt his dreams and send a shiver down his spine every time he heard a doorbell.

Walking into the crowded church more than a month after he was told his father was dead, that he wasn't coming home, holding his mother's shaking hand, Bryan was a very small boy in a throng of much bigger people. Sitting in the hard wooden pew, trapped below winter-clothed adults, all of them wearing black, Bryan cried because his mother cried. The shiny brown coffin carrying his father's remains was wheeled down the aisle. The boy caught only glimpses of the box between the legs of his father's pallbearers. Bryan stood to get a better look.

"What's in the box, Mom?" His clear voice startled the crowd. His mother hushed him to silence and pressed him to her trembling breast. Finally she managed to tell him it was his father.

Freeing himself from his mother's embrace and turning in the seat to better see where the men were taking the box, Bryan asked in a whisper that only his mother heard, "Why don't they let him out, Mom? How come I can't see him?"

Strange male hands reached down and tapped his head, his shoulders. Strongly scented women bent over him, all trying to comfort him. But the only comfort for the young Bryan came from the ever-smiling picture of his father at home in his mother's room.

Three years later, Bryan's mother had remarried and taken a new surname for her and her son. Eight years old and his father no more than a framed photograph on his bedroom dresser, Bryan hadn't rebelled against the new man in his life. He was never asked to call him Dad or even Father. His mother had taken the photograph of her dead husband from her bedroom and placed it in her son's room. She never told him why and Bryan never asked. When he got older he figured it out for himself.

Every question the maturing Bryan had asked about his father was always answered. He was denied nothing. His stepfather was a good man. Bryan learned not only to accept but to like the man, though Bryan never told him or even once showed him any affection.

The picture of his father in full military uniform, complete with SCREAMING EAGLE on the left shoulder, consumed him. It was all he had left of his father. For Bryan there had been no closure.

When Bryan was twelve years old, his mom showed him another picture. He had not seen it before. She told him she had been waiting for the right time to show it to him. It was a clear black-and-white aerial shot that had been clipped from a Canadian newspaper. It showed a clearing cut out of a timbered ridge reaching down to a long lake. Near the top of the wide clearing stood the bronze statues of a tall soldier, holding on either side of him the reaching hands of a child, a young boy and a girl. His mother told him it was Gander, Newfoundland, in eastern Canada. The place where his father's plane had gone down. The place where he and many others had died. From that day on Bryan connected his father's picture with his life, the Gander newspaper photo with his death.

Through high school Bryan excelled at geography, especially Canadian and Middle Eastern geography. He knew where his father had spent the last six months of his life. He also knew where he had spent the last few minutes of his life. His ambition was to visit both places.

Bryan took one year of university geography, with an interest in cartography, but changed his mind in his second year. He finally graduated with a degree in engineering science. Bryan had not considered the military as a career; he preferred the world of private business. His mom and the man whom he never called Dad supported him financially through his years in secondary school. When he graduated he didn't owe money to anyone.

It wasn't hard for Bryan to find employment. He had graduated at the top of his class. A Midwestern engineering firm hired him straight out of school. They designed and built huge river dams and high-rises and other buildings, as well as long, water-spanning bridges all over the world. This year they had landed a contract in Dubai, and as a very promising junior engineer, Bryan was asked to go along for a thirty-day stint. He had agreed.

Dubai was part of the United Arab Emirates, second-largest of the seven Emirates on the Arabian Peninsula and having the largest population. It boasted the second-largest free-standing hotel in the world, the Burj Al Arab. The hotel was also the most expensive. And now in 2002, Dubai was hosting several other massive construction projects. Bryan's company was involved in one of them.

Bryan knew where Dubai was. It was situated on the Persian Gulf side of the Arabian Peninsula. His flight involved a fuelling stop in Gander, Newfoundland, another one in Cologne, West Germany, and from there he would fly to Cairo, Egypt, before travelling east to Dubai. Bryan read his flight manifest over and over. He couldn't believe what was happening. He was flying the identical route, in reverse, his father had taken on the day he died!

*****

BRYAN WALKED WITH THE other passengers into the international section of the Gander airport. They had a two-hour-and-forty-five-minute layover. He hoped it would be long enough. The ceiling in the spacious lounge where most of the passengers milled about, some stretching, others reclining, was very high. One wall was almost entirely covered by a very colourful abstract painting. Bryan walked to the opposite wall and looked up to get a better look. He had never seen such a large painting. He picked up and scanned through a brochure that interpreted the mural. The painting was a full seventy-two feet long, he read, and was painted by Kenneth Lockhead, a Saskatchewan man, in 1959. It was first looked upon by royalty in June of that same year, when the queen of England herself, Elizabeth II, officially opened the new terminal.

Bryan tried to make sense of the painting. He couldn't, so he read more of the explanation in the brochure. The painting was supposed to evoke human feelings and understanding of one of man's greatest fulfilled dreams – flight. No aircraft was identified. The skilled painter had not wanted to date the painting, but leave it mysterious forever. Bryan noticed that one of the large, white birds taking flight in the centre of the mural bore a striking resemblance to the huge Concorde, but he also knew that neither the Concorde nor any aircraft of its calibre were flying any skies in 1959. The artist had done his work well. Bryan looked up at the mural again. He could see a young girl looking at a flower, an old man feeding a bird, a young man staring at him through binoculars, a boy holding up a blue egg.

Bryan looked away from the painting and walked toward a metal sculpture, "Birds of Welcome," in the centre of the lounge. A sign said rubbing the birds' heads would bring you good luck. Bryan figured he needed assurance as well as good luck right now. He rubbed the polished heads of the birds. Turning to walk away, he glanced up at the mural again. The boy with the binoculars stared down at him, following him, encouraging him.

Several of the passengers went through security and headed away to board waiting taxis for a quick look at the town of Gander. Bryan's passport and work visa passed scrutiny by a friendly Canadian customs officer. He submitted to a scan, and the Canadian wished him a pleasant visit and allowed him to leave the secured area. Bryan walked into the bar. He wasn't much of a drinking man, and it was still a bit early for him. He recognized several people from his flight at the bar with drinks in their hands. Bryan studied the many bottles on the wall behind the tall, well-dressed bartender. One in particular stood out: Gibson's Finest. Bryan hadn't known that his family name was on a Canadian whisky bottle. He ordered a Coke, no ice, drank it quickly, nervously, and left the bar.

He walked outside. The day was warm, pleasant. A soft wind was blowing. Bryan didn't get aboard a taxi right away. Now that he was actually here, he wasn't sure if he could go through with it. He knew all about the statue and the Silent Witness Memorial. Twice he walked past the same empty cab. The second time the cab driver didn't look his way.

Bryan slid into the back seat of the cab and said, "Can you take me to the Silent Witness Memorial, please?" before the cabbie had a chance to ask where he wanted to go.

"No problem, ol' buddy," came the reply with a thick accent.

When they arrived at the site, Bryan asked to be driven to the blue lake. He told the cabbie he would walk the rest of the way and asked him if he would wait for him there. The taxi driver repeated "No problem, ol' buddy," and Bryan stepped onto the shoreline of Gander Lake. He watched the cab make its way back up the hill.

Waves of clear water washed upon the rocky beach. The summer wind rustled the green trees all around. There was no other sound. Bryan stared out over the lake. The sun shone brightly and was already west of its meridian to his right. Bryan always liked to know his directions. That meant north was at his back. One article he had read said that if the Arrow Air had gotten a little farther south, she might have crashed into the Gander Lake. There may have been less fire, less destruction of the plane, possibly even less loss of life. No one would ever know if that theory was right. He wondered if his dad's body was still here on the ground when his mother had answered the door that morning. That day had always stuck with him. It was the first time he had ever heard his mother cry.

Bryan turned away from the lake and, looking up the slope of the green ridge, began walking up the gravelled road. When he reached the Silent Witness site, the cab driver was standing outside his car and smoking quietly. The two men nodded in greeting as Bryan continued his exploration. A narrow wooden bridge led him over a small stream gurgling below. Bryan hesitated halfway up the prepared walk and was about to turn back.

"'Twill be all right, ol' buddy," came from the man watching him. The cabbie knew without being told why his passenger was here. Without turning around, Bryan straightened his shoulders and walked higher.

He had always thought the statue would look like his dad when he actually saw it. It didn't. He suddenly wondered if there were other sons thinking the same thing. Looking up at the determined metal soldier, Bryan figured the bronze likeness was everybody's dad, and the two children next to him all the innocent everywhere, who would forever need protection. Ironically, the man holding the small, trusting hands was clearly a soldier, but he bore no arms. The two children beside him carried olive branches in their hands.

The olive branch was a symbol of peace the world over. As far back as the ancient Greeks, the simple olive tree represented peace; their mythology told them the Goddess Athena had planted the very first olive tree. In the Hebrew Bible of Genesis, an olive branch was carried by a white dove, signifying that their God had finally stopped the floods that had covered the land. "And the dove came in to him in the evening; and, lo, in her mouth was an olive leaf plucked off: so Noah knew that the waters were abated from off the earth." Even in ancient Rome, defeated armies carried olive branches as a sign of peace. The symbol lives on in our modern times. In 1775 the United States presented to Great Britain an "Olive Branch Petition" in hopes of achieving a peaceful end to their conflict for independence. The seal of the United Nations features olive branches. The great seal of the United States of America is dominated by a mighty eagle; grasped firmly in its right talon is an olive branch sprouting thirteen leaves and bearing thirteen olives.

The lifelike bronze sculptures stand securely on one of Newfoundland's firmest symbols: rock. It is the strata upon which the island sits, forever grounded, unyielding, common yet indomitable, safe, unmoving – the home no one ever really leaves.

* * * * *

DOUG SHEPPARD WAS THE mayor of Gander in 1985. The terrible crash and loss of life of so many young Americans in his town was perhaps the biggest defining event of his life. Doug was an unassuming, compassionate, sensitive, and well-liked man. He was born on Indian Island, one of several beautiful islands in Newfoundland's Notre Dame Bay, where a magnificent, red-skinned Native people had once lived.

Sheppard was one of Gander's most respected mayors. He served the town well for twelve years. In 1986, the first anniversary of the Arrow Air crash, a memorial to mark the tragedy was erected in the American state of Kentucky, home of the 101st Airborne Division. In a well-manicured bluegrass park, surrounded by trees and adjoining a running stream, complete with a burbling rock waterfall, today a lone copper sculpture of a soldier stands on a rugged outcropping of limestone. The very much alive soldier, Spec. 4 John Ronnegren, who had spent time in the Sinai with the 101st, willingly modelled for the statue. The copper soldier stands at ease and is wearing the dun and splotched-green uniform and company beret of the 101st desert campaign. His shirt sleeves are rolled up, baring muscular arms. His left hand is behind his back, and in his right he carries a deadly-looking assault rifle.

As mayor of Gander, Doug Sheppard was invited to attend

the opening ceremonies on September 28, 1986. He gladly accepted.

Laid horizontally on the lush, blue-green Kentucky grass, directly below the seven-foot statue, are five granite plaques that could have been hewn from any Newfoundland headland. Chiselled into the rock faces for all to see are the names of the 248 soldiers who died on the faraway slopes of Gander Lake. Ten miles south of the staring soldier is the home base of the 101st Airborne Division, Fort Campbell. Where the heavily trafficked Fort Campbell Boulevard and the Edward T. Breathitt Pennyrile Parkway intersect, the dedicated memorial to honour the fallen Eagles will forever stand. A symbolic Bradford pear tree was planted for each of the soldiers, a growing memory springing out of the lush earth. Soil from the Gander crash site was sprinkled at the base of each fruit tree. On that day Johnny Cash, balladeer of the common people, dressed in his most sombre black, recited his own patriotic poem, *Rugged Old Flag*.

Mayor Sheppard was impressed with the park. He was determined to have a similar memorial at the actual location of the tragedy on the forested ridge near his own town of Gander. Sheppard approached Steve Shields, the Hopkinsville sculptor who had taken great pains to fashion the soldier from molten copper. The two men talked in earnest for a while. Mayor Sheppard wanted Shields to construct a similar monument where the Arrow Air crashed, and Shields agreed.

Back home in Gander, Sheppard contacted Lorne Rostotski. Rostotski was born on the Saskatchewan prairies and now called Gander home. He loved the scenic beauty that he saw all around the island of Newfoundland. No one had a better eye for it than Rostotski. Today he is an accomplished artist and photographer. He had opened his first studio in Gander in 1972. Doug Sheppard and Rostotski were good friends. When Doug asked him if he would be interested in designing a memorial for the crash site, Lorne agreed wholeheartedly.

When they were finished, they learned that Steve Shields would not trust his delicate work to regular trucking companies. He and his wife, Karen, a high school principal, secured his treasured work in their own van and drove it to Gander themselves. It was erected where the Arrow Air went down, with the lone soldier and the two children facing south to the United States. The monument to the "Screaming Eagles" was on the site by 1990, in time for the fifth anniversary of the disaster.

Near the statues, on the same ground, stands another memorial to the fallen Eagles, a twenty-two-foot-high Cross made from the emergency exit doors of the Arrow Air DC-8. It is called the "Cross of Sacrifice." Inscribed into the Cross are the words RENDEZVOUS WITH DESTINY. All around the aluminum Cross are planted 256 native Newfoundland trees.

Here in this northern forest, where animals come and peer through the fringe of trees at dusk and sometimes cross over the hallowed ground, where night birds sound and geese herald each season, people from a foreign land rest in friendly soil. Gander, Newfoundland, Canada, and Hopkinsville, Kentucky, United States of America, are forever linked in a common sorrow.

* * * * *

BRYAN TURNED HIS BACK to the monument and stood for a while looking down over the patch of ground where his father had died. The lake looked bluer from up here. Again he wondered if the outcome would have been different had the Arrow Air made it to the lake. Would the cold, blue waters of this peaceful lake have taken his father's life just as surely as the ridge had done? It didn't matter.

Bryan Gibson had not shed a tear for his father, but today his troubled mind finally found peace. This place, where his father and so many of his comrades had fallen, had finally given the son a quiet acceptance, a soul-lifting closure. He turned to the sculptured metal figures again. The universal soldier-father, shielding, protecting, comforting the two nameless waifs, did what all the turmoil over his father had not done. Bryan wept. He wept for the father he had lost. He wept for the wonderful father he had gained, who had protected and stood by him all those years, never asking anything of the boy and man he called his son. It was time he called him Dad. Bryan was glad he had come to Gander.

He started walking down the hill toward the waiting cabbie. His step was light, his decision made, a lifelong burden lifted.

"I am ready now, old buddy," he called to the taxi driver.

Without missing a beat the smiling cabbie called back, "Good fer you, me son."

Oh, great, thought Bryan, *just when I thought I had it figured out!*

EPILOGUE

Once again I went back to the Silent Witness Memorial. I had completed the manuscript *Where Eagles Lie Fallen* but hadn't submitted it to my publisher. It was late June 2010. Bright new leaves rustled in the soft wind. Summer-forest smells filled my senses. The small brook beneath the memorial knoll had slowed to a pleasant, musical trickle. I wondered, as I walked up the greening slope, about everything I had written about the lives that had ended here. About the lives that had been so affected by this place. Had I done a good enough job of it? Should I even submit my work to be read by the public? At the top of the grassy hill where the tall soldier and the two children stood, the air was cooler. There was a new flag flying since my last visit. It was Gander's own, with its long-necked male goose flying across its centre, and snapping in the wind with the other three flags.

One of the small GFT planes flew in over the lake and lowered for a landing. It glided for a while as it neared, and with an almost respectful silence, the graceful Cessna slid down the sky. I sat on one of the benches. A young man and woman walked up from their vehicle and stayed awhile. We spoke quiet greetings before they left.

When I was alone the feelings of being unworthy to record such an event crowded my thoughts again. Had I done justice to the sad story? Did I express with accuracy in the pages of my work the feelings of those I had interviewed? Was the tale told well enough for those who could not speak for themselves, the truly silent witnesses of the Gander tragedy? Feeling humbled because of what I now knew about this place, and the trust they had placed in me to share their story, I walked down the slope just as quietly as I had done the first time I had come here in search of answers.

* * * * *

The Screaming Eagles' credo: *To crush its enemies by FALLING upon them like a thunderbolt from the skies.*

OF THE 248 SERVICEMEN who died on December 12, 1985, in Gander, Newfoundland, all but twelve were members of the 101st Airborne Division (Air Assault), most of whom were from the Third Battalion, 502nd Infantry; eleven were from other Forces Command units; and one was an agent from the Criminal Investigation Command (CID). Eight crew members of the Arrow Air chartered aircraft also died, bringing the total casualties to 256.

It was the deadliest plane crash in Canadian history. It was also the highest death toll on any single day for the U.S. Armed Forces since World War II, including combat losses.

It is the way of soldiers to die fighting. In the unspoken thoughts of all fighting men, they know death is always possible. They fight anyway. They fight for a way of life unequalled anywhere. For family. For freedom.

The American sailors who had lost their battle with the cold winter waters below the looming cliffs of Chambers Cove had died fighting. They fought the sea. They fought the cold of the unforgiving North Atlantic Ocean. They fought for their lives.

The courageous men and women who perished when the Arrow Air fell out of the sky at Gander died without being given the chance to fight at all. Such was their rendezvous with destiny.

There remains to be told here, literally, hundreds more stories. It is impossible to relate all of them. I hope the few that I have told are a reasonable representative of all the others.

If I have brought any closure or some small measure of comfort to anyone with connections to this event, then my work has fulfilled its intended objective. If I have caused you any added pain, then I have failed. Please forgive me.

Gander 9/11

CLINT COLLINS WAS SWEATING. He was sitting in the hot, narrow cab of a mechanical tree harvester near the top of a steep west-facing ridge. He referred to his cubicle as "The Greenhouse." With the hot sun pouring through three glaring panels of steel-framed safety glass, his cab was tropical. He looked at the thermometer hanging over his head. It read 36°C. The dozens of hydraulic lines running through the cab only magnified the heat. His machine was not equipped with air conditioning. A small electrical fan secured to one corner of the cab blew the hot air down his neck.

The joysticks mounted to his comfortable chair, one for each hand, never left his grasp. The left stick had a button for four of his fingers, the right the same, plus one extra for his thumb. Mounted in front of him was a computer with a small screen. It told him how to process the trees. After a twelve-hour shift his fingers were always cramped.

He tilted the left joystick sideways. The machine slewed around, the big boom with cutting head dangling, hanging down over the hillside. The harvester rocked a little before the heavy, swaying head settled. Clint pulled the throttle to idle. With his back to the sun, relief from the direct heat was instant. He was facing Gander. It was no more than seventeen or so miles away, he figured. He could see the white radar dome near the airstrip.

Clint thought of his paternal grandfather, as he always did every time he looked at the white building against the horizon.

"I painted that place, my son. Didn't know what it was – not really. Something fer findin' planes long before a feller could see 'em. That's what they told us." His grandfather's memories of building the town of Gander were always vivid, simple, and honest.

"It was some place to be, watching all them warplanes come to land, I'll tell ya. Young I was, then, didn't know much about painting anything. We didn't have much money to spare fer paint in Hare Bay. I smeared the windows with white paint. But I got it off. Well, most of it." Clint smiled, remembering his grandfather's laugh.

A plane came in low on the distant ridge. Clint could see it was a big one, white with dark markings. She was too far to tell which colour. The airliner banked as he watched. The sun glinted along what he knew were her windows. She lowered for her final approach, dropping as gracefully as a goose toward the Gander runway. The aircraft was almost down to the tree-line. It passed the circular radar dome and for an instant the building's round roof was hidden from Clint's view. Then the plane rushed past, its high tail fin cutting through the dark trees like a silent scythe.

Clint had started his shift early, just after daylight, trying to beat the heat of the day. It wasn't working. At just past 10:00

a.m., it was scorching in his cab. Reaching forward he released the front glass panel snaps and slid the entire overhead window along its well-worn track. Now the outside air rushed in. It was cooler than the cab air, but it brought something else in with it: blood-seeking blackflies, mosquitoes, stouts, and foraging hornets. A two-inch meshed steel grill protected him from errant branches and trees. Brushing the sweat and a few blackflies from his muscular arms, he opened the throttle on his machine and started to slew it back to work. He noticed another plane above the Gander skyline. Like the other one, this was also a big one.

Pushing both travel pedals with his feet, Clint rolled the Komatsu closer to the forest edge, her steel pads crawling over the harvested slash. For the next few hours, he worked steadily, cutting and felling the trees, de-limbing and cutting them into computer-instructed lengths and piling them in neat rows of logs. It was a constantly jolting and jostling task that required a lot of concentration. The bugs, the sweltering heat, and the dust from the trees made it hard for a man to focus. Each time the massive cutting head ran out over a downed tree, pieces of dusty bark and twigs flew through the air. The sun, slanting down through the forest canopy, lit up the dust motes in thick clouds. Most of the dust ended up inside the cab of the harvester. The machine was always coated inside and out with a fine film of dust. So was the operator.

Several times during his many turnings, Clint noticed planes in the distance. All of them were landing at the distant airport. *Gander must be having a special on aviation fuel today,* he thought. It was a common thing to see a few planes approach the airport in the run of a day, but to see so many in a couple of hours was not. They were all huge airliners, too. They all seemed to be coming in over the Atlantic, just a few miles away.

Clint stopped for a much-needed lunch. Swinging his machine away from the sun, he let the engine idle for a while to cool itself, then shut it down. The sudden silence was pleasing. It made his ears ring. He switched a lever to lock the safety and swung out of the cab.

He had turned to climb backwards down over the high metal track when a noise above his head startled him. Another airplane was thundering toward Gander. She was close enough for Clint to see the identification letters and numbers along her fuselage. It was foreign. Russian, or one of the Eastern European countries, he figured. For a moment she seemed to stall overhead. Then the plane lunged downward, leaving a vapour trail of high-octane jet fuel behind as she headed toward the Gander airfield.

Clint stepped to the ground and stretched his lean, six-foot frame as he usually did whenever he got out of the cramping quarters of the harvester's cab. Lunch can in hand, he walked into the shade of the forest. It came to his mind that the tree he was sitting under would not be here by the end of his shift. He had always loved the living forest, but he had also been involved in his family's logging and sawmilling business since school. It felt good to spend time in the dark, silent forest. It was his favourite hobby. He was used to working alone. Clint had always been a loner. He decided to take another look at this droke of trees before he cut them down.

Walking through the thick stand of timber, he weaved his way among the mature trees. His leathered feet sank into the yellow moss and he made no sound in his passing. The day was still young, the bright sun still barely topping the tall trees. Sunlight filtered its way down the canopy of black spruce and softened the shadows. Clint stopped and sat on a tree that had fallen many years since.

Along the forest floor shone great windows of light, complete with shadowy curtains for greater effect. A light hissing, whispering sound sighed away through the droke, rising and falling with a freshening wind. Looking up from his seat in the "aisle," he watched and listened to the free show as if for the first time. The sound of the wind through the thousands of treetops was hard to define. It was like an east wind blowing summer-soft up the blue bay where he lived. Now it was like the rushing sound of a river muted with distance. It was a glorious, lonely, wispy sound. It made him feel very alone and very small. As he watched, the heavy treetops high above swayed gently and splashed more sunlight down upon the waiting stage. The movement at the top was instantly brought to the forest floor, the undulating shadows causing the curtains to shimmer in the yellow window below.

His body cooled in this quiet place of rest and his mind returned to the work waiting behind him. Heading away from the scene, Clint stepped through the yellow window but never broke the mossy pane.

He heard another plane approaching as he sat with his back against a tree to eat his lunch. It was higher than the last one and seemed to be circling just outside the Gander airport. A black plume of smoke trailed behind it. This one was angled down for a Gander landing, too.

Clint opened a tin of salmon packed in water, his favourite.

A Thermos of juice and a large apple added to his usual fare. He seldom ate bread or sweet cookies. Sitting in the hot machine all day, the carbs made him lethargic and sleepy. He would have loved to have an air conditioner installed in his machine, but the $4,000 cost for one was more than their small company could afford right now.

A grey jay appeared. It was lean and dark. A young one, he knew. Others appeared; two were bigger, with distinctive grey and white feathers. The entire family was here, chirping and twitching, hopping from branch to low, trembling branch, waiting. Clint grinned and pulled out of his lunch tin a bag filled with bread crumbs he had brought from home. He was sure they knew he had brought food for them. Every day, as soon as he shut the engine down, they came around. They always followed him to where he ate, and he never disappointed them. He pulled an expensive camera out of its case and waited for the birds to approach the food he laid on a moss-covered blowdown. Photography was another of his favourite hobbies.

Clint focused his camera on the birds. Nothing out of the ordinary from the jays to photograph today, he decided. He suddenly sprang to his feet. Out of the ordinary! By God, what was happening in the skies over his head was anything but ordinary! He walked away from the coolness of the forest and stepped into the full glare of the sun. He faced Gander. Another plane was preparing to land, coming in from the ocean with a roar. High above that one he saw another glint of light. It was as if the aircraft above was waiting its turn to land. But what aroused his curiosity most was something else missing from the clear sky. The GFT planes were nowhere to be seen. The little twin-engine Cessnas of the Gander Flight Training school were always buzzing around the ridges, their aerial diving and circling a constant in the skies every day.

There was something else he hadn't noticed, either. He had not seen one plane leave the ground. How could that be? If all of the planes he had seen landing were still on the ground, the airfield must be filled. He watched as the plane that had been circling high above lowered itself to earth and disappeared below the treeline like the others. Still, not one plane was leaving Gander.

There had to be something wrong, he was sure of it. But what could it be? It was a perfectly sunny day; no storms of wind or heavy rain were even in the forecast. It just didn't make sense. He wished the radio in his machine were working. If there was something going on, the radio would report it, for sure. A low-hanging branch had ripped the radio antenna from the tree harvester's roof days ago. He didn't listen to the radio much, anyway. Taped music of his own choosing was his thing. Then his cellphone rang.

Even before he asked a question, the sound of his mother's frantic voice told him there was something terribly wrong. She was calling from St. John's. At first he thought his mother was calling about the thyroid surgery she was scheduled to have tomorrow. However, her own personal problems were not on her mind this day.

"There has been an attack on the Americans!" she told her son, her voice very loud and excited. "We are watching it on TV right now. Two planes have slammed into two different skyscrapers in New York City. The Twin Towers of the World Trade Center, they are called. The news is saying it was not an accident. It was intentional. They've cut the two buildings nearly in two. They are burning out of control."

"That must be what's behind all the planes I've seen landing at Gander all day," Clint interrupted.

"Oh yes, the planes are landing all over the island. They are not allowed to land anywhere in the States. There are reports of still more planes in the air ready to crash into more buildings. Even the White House may be next, they say! I can't believe what I'm seeing. Oh my God, all those poor innocent people." His mother's voice became choked and trembling.

Clint's father's voice came on the line. "Is everything all right where you are, my son?"

"The sky seems to be filled with planes headed for Gander. Apart from that it's okay here, Dad. What's happening?"

"I'm not sure what the hell is going on, my son. The TV is showing a war zone. New York City seems to be under siege. Someone is using airliners as bombs. The Pentagon has been bombed by another plane, they say. Another one has gone down somewhere in Pennsylvania. They're saying she may have been headed for the White House. All the planes were loaded with passengers.

"The States have shut down their airfields, even to their own planes. The skies here in St. John's are filled with circling planes."

His father paused. Clint could hear cries in the background.

"We are in a bank here on Elizabeth Avenue. We're watching the TV and they keep showing the planes crashing into the buildings. It's like nothing I've ever seen before. The planes are being used as weapons.

"People are crying here. Both of the buildings we are

watching are going to fall. There must be thousands of people in those towers . . . the ones above the fires will never get out." His father's voice trembled. He coughed, then spoke again. "Get out of the woods, my son. Go home to your family. We'll call again later. Bye, son."

"Bye, Dad." The phone went silent in Clint's hand.

He looked down the slope and west across the valley below to Gander. The forest had never seemed so quiet. Even the jays were silent. Then the drone of another plane came to his ears. As he watched it came into his view, low down and angled toward the airport town. He walked down the cutover toward the logging road.

The Reid brothers from Bishop's Falls were building an extension to the existing logging road. Seeing Clint approach them, they called out. They figured he had had trouble with his machine. They shut their equipment down and listened in disbelief as Clint told them about the phone call from his parents. Both of the Reid men had seen a few planes during the course of the day but hadn't paid much attention to them. They were at the bottom of a wooded valley and could see nothing above the narrow right-of-way jutting through the forest like a green tunnel.

The three men drove back out the road to their trailer setup. It was above the valley and had a clear view of the Gander ridge. Several planes were still circling. The men watched for a few minutes. Clint was right; not one plane could be seen rising from the airfield. What the three woodsmen didn't know was that Operation Yellow, the name given to the diverting of all aircraft heading for the United States onto Canadian soil, had begun.

One of the brothers started the generator. The three men hurried into the trailer and turned on the small television inside. They could only get two channels, but it was enough. There was no regular programming on this day; both channels were frantic with the ongoing news. All three men stared without saying a word at the horrific scenes displayed before them on the small screen.

Seasoned journalists were speaking in unnatural, uncharacteristically excited voices as they tried to describe, to their biggest audience ever, what was happening in America. The New York City landscape rising above the banks of the grey Hudson River was in peril. Two of its highest buildings were belching great plumes of dense, grey-black smoke out of their upper levels. Clint's mother had been right. America was under brutal attack – the weapons used: hijacked, passenger-filled jetliners. Again and again the film was repeated.

The first of the towers was hit by a speeding aircraft scything through the steel and concrete building's superstructure like a sudden, vicious chop from an angry Judo fighter. Flames erupted from the stricken tower and billowed out, as if a volcano had suddenly freed itself sideways from its mountain prison. This first attack was initially filmed by a tourist who, seeing the airliner crash into the World Trade Center, thought he was filming a flight suddenly gone wrong. This flight was dead straight on its predetermined course. There was no mistake. Hundreds of cameras below recorded this deadly piece of history. News helicopters flew as close as they dared to the inferno, capturing in movie form the world's first airliner attack.

The second attack was more spectacular than the first, for now professional camera people were already filming on site. A second airliner came into frame as the rest of the world watched in disbelief. But it was all too true. The second plane sped like a magnificent shark that had just flexed its great tail into the final swift turn for a bite of its doomed prey. This silvery, fish-like predator buried the length of its sleek, tubular body, complete with its screaming human cargo, into the very heart of its unsuspecting prey. Suddenly the whole watching world believed. America was under a devastating, unprecedented attack.

It was as if the reporters, who had now stepped onto a world stage, were trying to find words to describe a world suddenly gone mad. Even among the planet's storied warmongers, a terrible line had been crossed. Never again would even the few tenuous rules of war be observed. On this day the world stared in horror at the sadistic, malicious, premeditated slaughter of hundreds of innocent men, women, and children.

No conflict would ever be the same again. On this sunny September day, the idea of conflicting countries waging their fight across battlefields, soldier to soldier, had vanished. The cowardly act of terrorism, designed to attack not the fighting man but his family, was suddenly the modern way to wage war.

Clint walked quietly and unnoticed out of the trailer. Outside and away from the TV noise, the forest was quiet. There were no roaring planes overhead, at least for now. *Maybe they have all landed*, Clint thought. The images of horror splayed across the screen were made all the more poignant here in this peaceful place. The image that affected the tall woodsman most was in his mind's eye: hundreds of desperate trapped people. The cameras had zoomed upwards, dragging into clear view broken windows hundreds of feet in the air. As the horrified audience watched, great explosions erupted out of the lower floors. Billowing flames vied with black smoke bursting upwards.

What at first looked to millions of viewers like falling debris wasn't debris at all. It was the twisting, flailing bodies of humans who had jumped, speeding earthward to their terrible deaths. The unfeeling flames, the deadly, choking smoke, and merciless heat that is the terror of all humans had reached them.

Then the first of the towers collapsed. It was as if a creature of mythical proportions, no longer able to endure the misery of its burning lower body, had bowed its great head in defeat and finally given up the ghost. The building fell onto itself in a great spreading cloud, dousing its own fires with its crushing weight. The agony of many ended with the disintegration of this symbol of wealth and power in the United States.

Clint and the two workers mumbled goodbyes to each other. No shouted jokes, no discussion about their work, as is the custom of working men everywhere. They didn't even talk about all they had just witnessed. They just climbed aboard their dust-covered pickup trucks and drove away to home. Above them another airliner appeared and screamed over the green valley.

* * * * *

ON A WARM SEPTEMBER morning in 2001, the entire world was changed forever. Newfoundland was once again at the centre of all North Atlantic air travel. The first reason was the same as the old one had been: its geography. The second reason hadn't changed, either: war. A war unlike any the world had ever experienced. A war of terror.

September 11 was unusually hot all around the island of Newfoundland. The air was still. No wind blew. "Indian Summer" was on everyone's lips, although that autumn date was still ten days away. The temperature was nearing 30°C. The summer-like weather was only the beginning of many unusual happenings on this September day.

In the skies approaching North America, huge jetliners winged their way along, holding within their metal hulls thousands of lives. Many of them were living their last moments of sweet life and never knew it. Most of them would be shaken to the roots of their soul. All of their lives would be forever changed. None of the living would ever forget this day. Some of them already knew they were going to die.

In the annals of flight, this day would contain the longest page. Not even *Kitty Hawk* and the Wright Brothers would compare. Some people would curse the day of first flight leading to this one. Dates would be set by, before, and after September 11, 2001. The ninth hour of this ninth month would be forever infamous, worldwide.

Nine-eleven would evoke memories of world-changing events that had never been seen on such a global scale. Some would refuse to fly again. The innocent would never again feel safe from distant battles, or from conflict that had always been a world away.

* * * * *

THOUSANDS OF 9/11 WANDERERS found their way to the Silent Witness Memorial near the town of Gander. On this day the three flags were at half-staff. Here in the forest, where so many peacekeepers had lost their lives, their countrymen stepped lightly.

One well-dressed lady reached out reluctantly and touched the bronzed image of the young boy statue, only to pluck her hand away as if her fingertips had been burned, though the boy's fingers were icy cold. It frightened her. Melina was travelling alone on her return from England, a passenger on a flight from London to Chicago, and then with a connecting flight to her home in San Francisco. She was as surprised as the rest of the passengers when their flight had been diverted to Gander. She had never been here before, but she knew all about this place.

Weeping visibly and turning away from the others, Melina stepped to the grassy edge of the memorial site. Through misty eyes she looked out over the long, blue lake below. It looked so peaceful and quiet here now. It was hard to believe such a dreadful thing had happened here sixteen years ago. Standing close beside her, a tall, elderly gentleman stood with his right arm around the shaking shoulders of a woman. His wife, Melina figured. A pretty woman, too, from what she could see of her face. The other woman was sobbing, her eyes partially covered by a dainty white handkerchief. Melina wondered if she was crying because of what was happening back in the States, or, like herself, she had another memory of Gander, Newfoundland.

The news of a change in their flight had come at 38,000 feet over the North Atlantic. They were nearer to North

America than they were to the British Isles. Their flight was ordered to land at Gander International Airport. There was no compromise. Captain Mike Ballard had to tell the 198 passengers aboard United Flight 929 that there was a change in their flight plan. He didn't tell them the entire U.S. airspace was shut down to all air traffic.

He had taken enough fuel on board for the entire flight to the States. Having to land in Newfoundland meant his huge aircraft was too heavy to land safely. He would have to dump fuel over the Atlantic. Ballard was glad his aircraft was equipped to jettison excess fuel. Not all of them were. Planes without this mechanism installed would have to circle until enough fuel was burned to lighten the plane for safe landing.

Ballard had ordered his crew to start the procedure. Weights and volumes were calculated before tons of fuel were dumped into the skies over Newfoundland. It was always done at high altitudes; the jet fuel supposedly dissipated before it reached the earth below. On September 11, 2001, the largest volume of fuel ever sky-jettisoned came out of the aircraft flying high over the Canadian side of the North Atlantic.

Melina was travelling alone. She wondered if fate had anything to do with her being here in Gander. She had flown across to England several times over the last few years. Her beautiful sister, Laura, was married to an Englishman who was involved in the fashion industry, and Melina had indulged in the free tickets they always gave her. Not that she couldn't travel on her own dime; she was a fairly well-to-do woman. Still, her wealthy brother-in-law always refused Melina's offer to pay her own way. Melina loved visiting her only sibling.

She knew that a few planes from her country still refuelled

here in Gander. These were usually smaller airliners. Melina always purposely avoided them. She always booked her flight, well in advance, on the largest airplanes. She didn't want to land at Gander. She didn't want to put her feet on the ground that had so brutally claimed her only child.

Melina's memory of December 12, 1985, was not only a painful reminder of the day her soldier son had died. That day was also the beginning of a change in her life that even now, years later, she could hardly believe. It was about to become even more unbelievable.

Melina sat on one of the wooden benches and looked down over the green ridge to the calm waters of the lake below. The slight wind was sweet-scented and pleasant on her face. She had stopped crying. This was not an ugly place rooted out of the woods as if by some monstrous, ground-foraging beast. The smudged black-and-white newspaper photos of the Arrow Air crash had misinformed her.

Melina fully understood the carnage that had taken place here on that early December morning sixteen years ago. But now, for the first time after all those years, she saw something different here. The unopened coffin, blanketed with the flag that Michael had died for, had always tormented her. Had her son really been inside that draped casket? If he was, why wasn't she allowed to look at his sweet face one last time? She had asked for that privilege, as did hundreds of others, but it was of no use. Her son was laid to rest sight unseen.

She suddenly realized what had bothered her so much about coming here. It wasn't the fear of seeing where Michael had died. It was the reluctant acceptance that her son was still here. She was suddenly glad she was here on this sacred

WHERE EAGLES LIE FALLEN

ground. When she learned that the people of Gander had arranged for tours to this site, she had at first refused to come here. It had taken great resolve and more fortitude than she knew she possessed to finally agree to the short bus ride to the crash site.

Turning off the Trans-Canada Highway, she saw a sign with the familiar 101st Screaming Eagles emblem. SILENT WITNESS MEMORIAL. Melina was the last one off the bus. It was more than not knowing what she would see on the cleared space beyond the dirt road where the bus was parked that held her back. It was the fear of facing head-on her own private dread she had kept hidden all these years. She had never believed Michael was beneath the ground where she carried flowers each December 12. It had always eaten away at her to pretend with so many others that Michael was there.

Finally, Melina had a closure she had never known was possible. Until now, her Michael had simply gone away and never come home. His young life had ended here on the slope of this valley. He had gone away with the boys but was resting here with the men. It was enough. Her mind filling with distant thoughts, Melina sat quietly above the Gander Lake and let her memory flood her senses.

* * * * *

AT ONLY SEVENTEEN YEARS of age, Michael already had his mind made up. He was going to make the military his life. And not just any branch of it. He wanted nothing more than to be a part of the "Screaming Eagles." His mother was against it for a while. There was so much violence going on in the world these days, she told him, and their country was involved in most of it. She didn't see why he had to go "marching off to war," to which Michael replied in his always jaunty way, "No one 'marches' off to war anymore, Mother. They all go by huge ships or fast planes."

They had both laughed then. All of their disagreements ended that way. When they had told her husband, Martin, he was very pleased with his son's decision. "It is a career move for sure," he said. "Why shouldn't Mike go? He's healthy enough. Besides, a man couldn't do a better thing for his country."

Martin had never been in any of his country's Armed Forces. He had even gotten past his draft notice of the '60s. He had "fallen arches" – flat feet. He had always been sensitive, and even bitter about it. Melina couldn't see any difference between Martin's feet than anyone else's. Her husband always steered any conversation away from the U.S. Armed Forces.

Michael was so determined to succeed in the Army, he didn't even complain about boot camp. He studied the history of the 101st brigade. He knew every battle the group had fought in. He paid special attention to the battles involving parachute drops: the Airborne Division – his favourite.

Melina would never forget her son's excited voice the day he called to tell her his battalion was assigned to the MFO in the Middle East. It was early spring. They would ship out to the Sinai in the coming summer, Michael had told her. He was as excited as a schoolboy who had just graduated with honours.

When she told her husband, he was furious. Melina had never seen him like this before. Martin raved on and on about his son going "over there" among the "damn Arabs." To Martin, anywhere in the Middle East meant Arabs.

She tried to reason with her husband. "It's a peacekeeping mission," she said. "Michael is not going to war. He sees it as a great opportunity to see the world. Besides, Michael always loved the warm weather." Melina tried her best to make light of the situation. It was no use. Martin just continued on with his tirade about the Arabs and the United States government wasting money to protect them. Strangely enough, Martin never said a word about his displeasure to his son.

The next few months brought to Melina wonders of a world that she would probably never see. Michael wrote descriptive letters of the geography and the thrill of experiencing a different culture. He sent beautiful pictures from his station in the south camp: of brilliant sunsets splashing across the Gulf of Suez, of sunrises blazing across the hot sands of Saudi Arabia and bringing their morning colours to the Red Sea, and of strange, high-prowed, single-masted fishing sloops heading out over a calm sea to catch fish with names she had never heard of. And people pictures, some of them with a smiling Michael posing with his comrades.

There were colourful pictures of busy souks where vendors hawked their wares. Crowded streets filled with dark-haired, dark-skinned people. *Michael could fit right in with them*, Melina thought. He was as dark as any of the men in the photos.

When Melina mentioned this to Martin, he flew into a terrible rage that frightened her. "No son of mine looks like a goddamn Arab," he shouted.

Melina tried to reason with him. "What's got into you, Martin? I only meant Michael has a dark complexion. He probably gets it from my Italian ancestry. I've always been tawny-skinned and you have a darkish, swarthy colour yourself," she said, trying to lighten the mood.

"Yeah, well maybe I have some of your Italian blood in me too," he said.

Martin had stormed out the door that day. When he returned later he had settled down, but he still wasn't himself. They had never spoken about the incident again. It was one of the things Melina hated about their marriage. Martin would never discuss any of their differences after the initial argument. She never knew what he kept festering inside.

That summer, Martin followed the government news more than he ever had, especially the military updates on America's peacekeeping role in the Middle East. The pictures on the TV screen always made him angry.

"Why the hell are they babysitting a bunch of dress-wearing, woman-beating Arabs anyway? Why would anyone want to live in that damn desert? If I had known Mike was headed for that Arab desert, I would never have agreed for him to join the Army." The only time Melina ever heard her husband curse was when he was referring to the Middle East and the Arabs. She didn't give it much further thought.

December 12, 1985, had not started as just another pre-Christmas day for Melina. She was up before the dawn. It was raw, windy. They didn't call Chicago the windy city for nothing. She was too excited to sleep. Michael was coming home today. Well, he would be at the Army base in Fort Campbell, Kentucky, today. He wouldn't get here in Chicago until the fourteenth. Still, just knowing her son was on home soil, only a state to the south away, filled her with a mother's joy.

Their house was in full Christmas swing. Inside and out,

strings of coloured lights festooned the walls and eaves. A ceiling-high natural fir tree, bought at a lot downtown, dominated the gaily lit home. The sweet-smelling tree filled the large living-room window. Melina wondered if she should take the goose out of the freezer. She was going to cook it the day Michael got home. She also had another one for Christmas Day. They rarely cooked turkey for Christmas. Michael loved succulent, slow-roasted geese, and not the store-bought ones, either. He preferred the wild birds shot just north of here, which were sold, cleaned and delicious, out of farmland sheds by the hunter-storytellers themselves.

Then the phone rang and shattered her plans and stopped her son from ever coming home. That phone call changed her life and her world forever. Michael was aboard the Arrow Air, which crashed barely past the takeoff runway, in Gander, Newfoundland. There were no survivors.

She waited in a semi-daze of shock and disbelief for Martin to come home from work. He had said nothing when she had phoned him. He said nothing later, standing in the centre of the open door, lunch tin in hand, his coat neatly buttoned to the chin. Sobbing aloud, she ran to him for comfort, wanting to feel his arms around her shaking shoulders, needing his strength. She got neither. Martin stood unmoving, staring at the silent TV. Melina had not turned it on all day. She didn't want Michael's death confirmed. She didn't want to see pictures of a plane crumpled onto a frozen strange land with her boy's body among the smouldering wreckage.

Martin turned on the TV. He sat on their cloth-covered couch still wearing his work boots and hat, his coat still tight to his chin. The screen filled with the excited reports of a military plane down with no survivors in Gander, Newfoundland. The first words that spat from between Martin's fiercely clenched teeth were, "Goddamn Arabs."

Melina had heard him say it dozens of times before, but never with such vindictiveness. It was an evil-sounding phrase with no room for remorse for the son who would never come home. It would only get worse, all through the next few terrible days of coming to grips with the death of her beloved Michael. The phone calls from friends and family. The constant reliving of the day of the crash at every explanation, until Melina felt she could not go on. The TV scenes that Martin was glued to and which she tried to avoid, of coffins waiting in dreary warehouses. The funerals she finally forced herself to watch on TV. The sobbing mothers and daughters and sisters, fathers and brothers. The screams of one young woman holding the hands of two small, blond-headed boys. The endless carefully prepared speeches. The braided formalities of a grieving military. The haunting, echoing moan of the *Last Post* that ended it all for some, but not for her.

January came and went without Michael's body coming home. Through it all Martin had shown no remorse. He had not shed a tear, none that she had seen, anyway. He bristled visibly when government officials were seen reaching for grieving hands in a gesture of condolence.

Melina had always suffered from the winter blahs. However, the winter of 1986 was the worst she had ever encountered. She was treated for depression, for grief-induced suffering. And still, Michael's wasn't one of the bodies that came. She called every day. Some days someone talked to her. Most days they didn't. March came and the last eight victims of the Arrow Air crash were finally released and shipped from the Dover Military Mortuary in the state of Delaware. Michael's body was one of them.

Melina knew Michael was not inside the flag-shrouded coffin. She suffered through the funeral service as if she were detached from herself. As if she were watching from a distance, going through the necessary motions. It was held in a church they never attended. Neither she nor Martin were churchgoers. She felt cold and out of place. Halfway through the service she suddenly wished she were a part of a church, any church. The singing brought her a peace she had not felt since Michael died. She was surprised by the feeling. Looking around she saw friends and even strangers crying. It made her weep. Martin stood by her side, but yet apart, dry-eyed and unfeeling. She wept as the empty box was lowered down.

Before spring ended Melina was diagnosed with SAD – seasonal affective disorder. They had named it right, she thought. She was sad all the time these days. She didn't think the seasons had anything to do with her depression.

Then Martin lost his job. He never told her why, exactly. She learned from one of his friends that his employer had had enough of his on-the-job tirades against the government, and his constant racist remarks, all against the Arabs and the Middle East. The owner and boss of the plant where he worked was originally from Lebanon.

With Martin out of work and at home, they fought almost all the time. It was always about the same thing. He was convinced that the Arrow Air plane was bombed by "them goddamn Arabs." Even before the possibility arose that it could be true, Martin was sure of it. He spent money they didn't have, travelling all over the place and meeting people who thought as he did. He refused to discuss the theory of ice on the aircraft's wings, or any other possible explanation for the plane crash. The preliminary Canadian and American government investigations into the fatal crash were nothing more than a "goddamn cover-up." If she felt so bad about Mike's death, why wouldn't she help him dig out the truth from the government? His son was killed by the "goddamn Arabs." Martin became impossible to live with. It was more than the loss of their son. He was obsessed with what he considered U.S. treachery. The Canadians were just as bad, going along with the Americans.

Their friends stopped coming over. Martin became a bitter man, old before his time. Worst of all, Melina didn't love him anymore. Before the long days of warm summer wind came in from Lake Michigan, Melina's favourite time, they were divorced. It had come to a terrible head one day in late March when Melina was spring cleaning. She was sorting through Michael's things, remembering his face that was so much like her own. Displayed on the kitchen table were dozens of the photographs he had sent from the Sinai.

Martin walked through the kitchen and, upon seeing the snapshots, grabbed a handful and began tearing them to shreds. Melina screamed at him and wrenched the precious paper from his grasp. Looking at the tattered reminders of her only son, she slapped her husband's disturbed face over and over. She only stopped when she was too weak to go on, and still he stood there, his face as red as a new sunburn from her furious slaps.

It was only a matter of time before they were separated. Melina was amazed to learn she was glad it was over. The last few months with Martin had drained from her any previous love she had had for him. He simply left after signing the divorce papers, gave her the house, kept the car, and took two suitcases filled with his clothes and nothing else. She had asked him, "Where will you go? What will you do?" Martin mumbled a reply, as if he never wanted to speak to her again. "I'm going to New York. I already have a job at the WTC there." Melina had never heard of it. And without saying another word, Martin drove out of her life. Melina never heard from him again.

Melina didn't have much of a social life anymore, although her friends still called and made regular visits. She always looked forward to it; she had always loved being with her friends. She even started going to church. The quiet time there seemed to soothe her grieving. She never went out to movies or restaurants, and only rarely did she go for coffee, although her friends always asked her to go with them. She wouldn't hear talk of meeting another man. Several men called her, most of whom she knew, all very polite, but she had made up her mind that she would never get involved with a man again.

Then she met Myron.

It was December 12, 1986. Melina drove to Fort Campbell for the very first anniversary of Michael's death. A memorial service was being held there. Although she didn't feel the need to go for herself, she just felt Michael should be represented. She couldn't let her boy down.

The memorial service was just as emotional as the funeral had been. Melina found the music more heart-rending than any of the words spoken. She had discovered since she started attending church services that music reached her very soul and aroused her deepest emotions. She looked around several times half expecting to see Martin. He never came.

They were all standing. A choir was singing a hymn that everyone else seemed to know. Melina only knew the tune. She strained to hear the lyrics as the wisdom of the poet's work blended with the sweet music, moving her to tears. She stumbled slightly. A man standing next to her gently took her elbow, steadying her. Melina looked up and through her misting eyes saw the deepest pair of blue eyes filled with compassion. Neither of them spoke.

Everyone walked to a building nearby for refreshments after the service. Melina sensed the man's presence behind her. She poured her coffee from the urn and, with a blueberry muffin in her other hand, she looked around for a place to sit.

"You can join me if you'd like," he said. It was the man with the blue eyes. Melina felt awkward, but the coffee was hot in her hand.

"Thank you," she said. She sat down opposite the man.

"A lot of hurtin' here today," he said. "Did you lose your husband?"

"I – er, no. Michael – er – my son, sorry," Melina stammered.

"It's all right. I'm Myron." He held out his hand. Melina reached across the table. His grip was firm, gentle, very warm, and somehow reassuring. An odd name for a man, she thought.

"Did you lose a son, too?" she asked. She had gained control of her voice.

"No, my only brother," said Myron.

From that first emotional encounter, their relationship grew into something more meaningful than Melina thought possible. They were in love from the start. They met openly whenever Myron was in Chicago, but it wasn't enough. On February 14, 1987, the happiest Valentine's Day of Melina's life, Myron asked her to marry him. Staring into the blue eyes she had come to love so much, Melina happily said yes.

When she told her friends she was going to marry Myron, they were overjoyed for her. Everyone loved Myron. They teased her about all the names beginning with the letter M in her life. Melina hadn't noticed. When she did consider it, it dawned on her that the letter M was the thirteenth letter of the alphabet. Not a lucky number. Still, Myron was the third letter M in her life – besides her own – and everyone considered three a lucky number. She shrugged the thought aside. Melina was not a superstitious woman.

Myron was a successful broker for several businesses, one of which Melina didn't even know existed. Expensive yachts were transported all over the world by ship – not for sale, but by their owners. People with lots of money wanted their luxury boats to be wherever they chose to stay for extended periods of time. Most of this business carried boats from the eastern seaboard of North America, including Canada, through the Panama Canal to the warm Pacific winters. When summer returned north the boats followed. Some of them followed the sun all year round. They wanted their boats, but they didn't want to sail them all over the world. Myron saw to it the custom-built yachts accompanied their owners.

Myron lived in San Francisco. A great deal of his work took him to the New York City area. He drove across the country several times a year. Sometimes he rode by train. He seldom flew. Myron had a fear of flying.

*****

THEY WERE MARRIED ON a warm spring day in 1987. Neither of them saw any reason to delay the marriage. They were married in the same church she had gone to for Michael's funeral. For some unexplained reason deep within her, it was what she wanted. It was the last place she had had any contact with her son. That day the church witnessed her terrible pain. Today it saw her joy. She believed Michael would have approved.

Selling her house was a breeze. Myron was involved in the real estate business, too. Melina's house fetched a price she never expected. The agent's fees were waived. There was no mortgage on the house. The money from the sale was her own. It was as easy as that. She was going to live in San Francisco. She couldn't believe things could be so simple, but Myron had all the right contacts. All she had to do was clean out the place.

Melina offered some of her clothes, kitchenware, and other items to her friends. Most of it she gave to a very grateful church. She enjoyed being associated with the church now. In the basement she went through a few tools Martin had left behind. It was the kind of stuff you could find at any yard sale, she thought. She gave it all away. In the back of a set of plywood cupboards Martin had screwed to the wall, she found a small, heavy cardboard box. It was sealed with a wide grey band of duct tape. There were no markings of any kind on the box. It wasn't hers. With a pair of scissors, Melina cut the sticky tape and opened the box lid. She glanced quickly around the basement, as if she were uncovering some secret find that no one else should see.

From outside came the sound of children playing, a reassuring, pleasant, everyday sound. A car horn blared once.

Still, Melina felt she was involved in some clandestine act, uncovering some secret that Martin had hidden from her. She wished he had taken the box with him. If its contents were sensitive enough to her former husband to keep hidden, why hadn't he taken it with him? It was as if Martin had wanted her to find this box, but why? And if that were so, why had he hidden it from her? She already hated what the box would reveal.

Inside the box and sorted in Martin's methodical manner was the fragmented history of her former husband. All correspondences were sorted according to date. The document on top of the neatly stacked papers, most of them personal letters, had the oldest date. It was from an out-of-state orphanage. Apparently Martin had been searching for his own roots.

Melina quickly rifled through the contents and saw that the information had taken Martin years to gather. The first date was 1964. Melina suddenly realized it was only two months before Michael was born. It appeared that knowing he was going to be a parent had started Martin on a search for his own parents.

All Martin had ever told Melina about his life was that he had been an orphan. It was obvious he hadn't wanted to talk about it and she hadn't pursued the matter any further. Michael was simply told that his paternal grandparents had died long ago. Melina's own parents lived clear across the country in Washington state. Her father made a meagre living as a fisherman on the American side of the Juan de Fuca Strait, which separates Canada from the United States. Her parents seldom had money to travel across the States to visit. She and Martin didn't, either. Michael hadn't had any grandparenting in his life.

Melina took the cardboard box under her arm and headed upstairs. She no longer felt guilty for looking at Martin's personal effects. This had been intentionally left for her to see. She was sure of it. Settled at her kitchen table facing the quiet mid-morning street, she began to examine her find.

Martin had spent seven years in three orphanages. He was trying to find out where. From the last of the orphanages, in Illinois, he had started his young life in no less than five foster homes. Martin had kept a meticulous record of all his contacts, including the postage dates. Three of the foster homes he had contacted didn't answer. One wrote back to say they were no longer in that business and had kept no records. The last one, a lady – Martin had fond memories of living in her home – wrote to tell him she was no longer with homecare. She said she often thought of him, and wished him well in his search. The lady also told him the name and address of one of the foster homes where he had been. It was Martin's first lead.

The letter from that home told him they were not permitted to give out information about their clients. It only made Martin more determined; it meant they had something to hide. His letters continued to go unanswered. A few were to government agencies. He was told the law didn't allow state homes to divulge sensitive information. Years went by. The old, faded envelopes were stacked atop new ones. Then the law changed. Martin was finally allowed to see his personal file. It was not pretty.

His mother's name was Michelle. The name jumped off the page to Melina's startled mind. Another "M" name in her life. But more than that. She remembered how Martin had insisted on the name Michael for their son. She had agreed because she liked the name. She wondered now if as a child Martin had been told his mother's name a few times and had kept it in his subconscious all these years. Melina had no answers for it. How could Martin have kept such a thing to himself? Michelle was Michael's grandmother. Michael had a right to know. Melina was furious with her former husband. Then she picked up the copies of several pages of a police report written in a poor longhand style.

The writer appeared to be in a hurry, the lines scrawled and jumbled. Halfway through the document, Melina's gasp was only partly silenced by a trembling hand covering her mouth. Before she finished, Melina was sobbing aloud, shattering the silence in her quiet kitchen.

The policeman who had written the detailed report did it very well before he had signed it. It was also signed by Michelle. The signature was light, the ink laid gently down by a weak hand. Michelle had given no surname. No identification was found on her person. She had no personal effects. Her statement was given from an intensive care hospital bed. There was a tube in both her nostrils, an intravenous in her arm. A thin wire attached to her right forefinger trailed from a machine with a constantly jumping graph on its output screen. She was barely awake. She was Caucasian and blonde with green eyes, about five foot four, and weighed an estimated 130 pounds. She was once very attractive. She was barely seventeen.

Michelle had just given birth to a baby. It was a boy. The dark-haired baby, there in the hospital, was doing just fine. His mother was not. She had given birth to her baby not in the hospital, but in a very cheap, dirty rooming house in Cairo, Illinois. There was no view of the slow-moving Ohio River as advertised by the half-lit window sign. On a single upstairs bed that needed cleaning, she had laboured in agony for hours. Her desperate cries finally brought a scared young woman knocking on her door. When the door was opened, Michelle was splayed naked in the dingy hall light. Between her spread legs was a screaming baby boy. He was face down, his tiny, desperate hands and feet painting the dirty counterpane from the spreading pool of his mother's blood.

The blaring ambulance took her and her baby away from the high-banked river and saved her life – for a while.

Melina put the paper down, pushed herself away from the table, and stood staring out the window. Two robins, the first she had seen this year, pranced and ran across her greening lawn. Melina burst into tears. That poor woman, alone at a time when no woman should ever be alone. And Martin born under these conditions. Funny, she could never visualize Martin as a baby. She had never seen a baby picture of him. She doubted if any existed. She returned to the table, hating to read the morbid drama, yet unable to resist.

* * * * *

BARELY A YEAR BEFORE, Michelle had visited the Middle East. She told the scribbling policeman her father was a wealthy and very influential man. She wouldn't tell his name. He had gladly paid for his daughter's first major solo trip abroad. He knew she was a geography buff who loved to travel with her parents, especially to exotic places. It would be a high school graduation gift. He wanted to book her in the finest hotels in Egypt and Israel, with secured guided tours. It wasn't what Michelle wanted. She wanted to experience the simple way of living in the East, not the façade prepared for wealthy American clients. They argued. She won, as she always did. She went with another, older friend of hers who had been there before. Michelle only said her name was Louise.

For several weeks she had enjoyed the mysteries of the Eastern Mediterranean. Egypt and Israel and Lebanon and even Kuwait. Michelle wanted more. She wanted to go to the place touted as the very "cradle of humanity." Ancient Persia itself: Iraq. They couldn't get entry permits. She, Michelle, was too young to be travelling without a male escort. She was infuriated but could do nothing about it.

Aside from that Michelle was having the time of her life. She had seen a part of the world that she had read about and studied in school. Michelle and her friend had sought out and found sights away from the beaten path. She had found the people of Saudi Arabia just as friendly as any of the other countries she had visited. Knowing her limits as a woman in a male-dominated country infuriated her, but it kept her safe. She hated the way the women dressed and believed it wasn't their choice. She had purchased a burka, complete with hijab, and even bought a niqab, the clinging veil that hid a woman's face, to complete her Middle Eastern costume. The one time she had worn it was her last. She had never felt so uncomfortable in her life. The heat under the clothes soaked her skin. The lack of freedom choked her.

The women she had talked to in her travels had never heard of women having the same rights as men. She was determined to have her say to a group of these bearded male chauvinists. She should have kept her western thoughts to herself.

* * * * *

WHEN MICHELLE TOLD HER father she was pregnant, he was outraged. When he later learned she had been brutally raped in Saudi Arabia, he became incensed. He now knew his daughter's first story of falling from a motorbike while on her vacation had been a horrendous lie. Michelle's father knew enough about world politics to realize that the law could not help. Besides, he could never handle the scandal such an international investigation would cause.

He demanded Michelle have an immediate abortion. He didn't care that she was more than six months pregnant. When she refused, saying it was her baby, it was innocent of its conception, he ordered her out of his house. Her mother wasn't much help; she was so caught up in the opulent trappings of her fickle society that even the stricken look on her young daughter's face couldn't reach her.

Michelle's words recorded by the policeman were her last. She died one hour later of complications due to a very difficult and unsupervised childbirth. She never saw her dark-skinned boy.

Melina was weeping uncontrollably. The last few hours of this woman's life, shared with a policeman, read like a cheap novel. She now knew why Martin had cursed every time the Middle East was mentioned. The photographs of Michael associated with these people he hated so much. The very picture of his dark-skinned son blending with the others. He hated his

Arabian blood. Melina looked at the date on the last envelope that had revealed the last page of her former husband's troubled life. It was the same day they had had the fight. She would never forget that day or the date, when she had hit him in the face repeatedly.

Melina felt drained. She wished she had never seen the box of hidden secrets. She had always wondered how Michael would have reacted if he had known his father's secret. Would he have still smiled his wonderful smile among the Arabs?

* * * * *

THE BLARING HORN SOUNDED from the bus waiting to take her back to Gander, bringing Melina out of her deep, troubled thoughts. This place where her son rested had awakened in her a dark secret that she had kept hidden from everyone. She had not told Myron. She suddenly had the strange feeling that by reliving the secret here, she had shared it with her son. She would share her secret with Myron as soon as she got back to California.

The range of emotions that had invaded Melina's mind were soon allayed by the wonderful people of Gander. Melina was amazed to learn that not only was this amazing outpouring of hospitality coming from this small town itself, but from other towns miles away. They told her in kilometres. It took her awhile to convert it to miles. Some of the places were called outports. Melina had never heard the word before. Outports were small towns alongside of the ocean, she was told.

All of the airliners and private jets alike had been directed out of the sky to airports all over Canada, most of them to

Newfoundland. From as far north as Labrador, they came out of the sky to the safety of Canadian soil. At 10:10 a.m. on that morning, the first of thirty-nine planes landed in Gander. They were not allowed to leave. They would be stranded here for days.

The passengers swelled the airport town's population of around 10,000 to nearly 17,000 in a matter of a few hours. The hotels, motels, and B & B's were filled to their maximum capacity. The call went out to the public for assistance. The citizens of Gander and its satellite towns came willingly to their aid. Overnight it seemed every public building in town was made available to "the plane people." Truckloads of food appeared. Cots were set up in school gyms. Every service club was open twenty-four hours a day. Everyone volunteered for anything they could do. All for the exclusive use of a people the Newfoundlanders had never seen, and all of it was free.

This simple kindness dulled the pain of her experience at the Silent Witness Memorial. Everyone spoke to the stranded passengers. They were all strangers, but Melina could tell all of the local people by their accent. She loved it, although she had to listen carefully when they talked to her, they talked so fast. She was amazed at the way everyone greeted her. No one passed without saying hello. Even the quick nod of their heads in passing was a form of greeting here. Many of the Newfoundlanders greeted them with a quick, "'Ow ya gettin' on," the salutation rolling off their tongues, sincere and genuine. Even when she asked what it meant, when she said, "How are you getting on," it didn't sound quite right.

The people of Gander not only offered their help at the different charitable venues in their town, they open up their homes. Men, women, and children were offered shelter, showers, home-cooked meals. The residents were willing to sleep on their sofas, giving up their own comfortable beds to strangers who would become friends. Never before had such a small area poured out, for all the world to see, the love of pure human compassion as did these simple people of an island commonly fraught with countless sorrows of its own.

Along with thousands of others who had landed here in Canada so unexpectedly, Melina felt helpless. Her country was being besieged by as yet an unknown people. No one could believe the senseless attacks that had been perpetrated on ordinary citizens. Everyone was asking why. Every TV and car radio blared out in repetition the historic events of this day. Never before had such a one-sided battle been witnessed on a world stage. No one, anywhere, would ever feel completely safe again.

Melina couldn't look anymore at the scenes replayed over and over in the school gym. Again and again the dramatic film was played, the two airliners transformed into missiles, slamming into the two towers like Japanese kamikazes. Again and again she saw the buildings collapse, until it didn't look real anymore. She was turning to leave the crowded building when several people gasped in unison. New film was being shown. Melina turned toward the screen with the others, unable to resist the human need to know.

No sound came from the gathered audience save for the loud indrawn breaths of disbelief. The cruel modern camera brought to reality what everyone watching didn't want to see from one of the towers high above the streets of New York. Just above the searing inferno and ever visible in large broken windows through the swirling, deadly black smoke, were hundreds of trapped and terrified people. Something fell from one of the windows.

"A desk," one of the watchers said quietly.

"No, it's a man," another hissed.

Melina's knees gave out and her stomach churned. She staggered toward the door, bent and gasping for breath. One of the local women, wearing what Melina knew was a homemade, full-length, colourful apron, came to her aid.

"Can I help you, my love? I know 'tis hard to watch. It's unbelievable what is happening, isn't it?" Melina only nodded, standing straight as the fresh air came to her. Thanking the apron-clad lady for her concern, she walked away.

The three letters shouted by an anxious newsman were what had seared the memory of Martin into her brain. They sounded as she had heard them years ago. The day Martin had walked out of her life, he had told her he had a job in New York working in the WTC. She had not connected the three letters as an acronym for the World Trade Center until the reporter had just said them. Melina was shaken to her very roots. For her this day of horror had finally been made personal. She couldn't believe this was all happening to her.

Melina's day was far from over. It would get more personal.

* * * * *

THE WOMAN WHO HAD come to Melina's aid before saw her re-enter the gym.

"Are you all right, my love? Can I get anything for you?" The woman's voice was soft, sincere, her concern genuine.

"No, I – I'll be okay. It's just this day. I can't – I – thank you again for your concern." Melina's voice was faltering.

She sat at a table. The aproned lady brought her coffee and sat beside her. They talked for a while, about their different homes, the weather, the crowds around town. But Melina never told her that she believed her former husband had been in the burning towers. They shared names. There is a way that only women have of reaching out to sorrow, an inbred mothering instinct that stands strong and true, sharing the burden of grieving, easing the hard times. It is unique to their gender the world over. Melina would always remember this wonderful woman as the lady in the apron.

Melina felt better. The woman's way of talking calmed her. She suddenly wanted more than anything else to hear Myron's voice. The last time she had talked to him, she had been in England. He was leaving Newark today to meet her in San Francisco. He had driven east to attend to his usual beginning-of-autumn business. He was flying back to California rather than drive back, he had told her. He would fight his fear of flying just to be there when she got home.

Melina asked her friend if the phones were free.

"No, my love, there's a lineup a mile long. But I'll watch for a break and let you know as soon as one becomes available." And, smiling her always present smile, the "apron lady" walked away.

Dinnertime came – the Newfoundlanders called it supper – all served by an army of volunteers who seemed to be tireless. Melina's apron lady paid extra attention to her. She had a friend here now. It made all the difference.

The excellent meal was over. People were milling around the TVs watching the endless filming of carnage. Now the footage showed the plane that had crashed into an empty field. They were saying this one had been headed for the White House in Washington D.C.

Melina saw her friend hurrying toward her. She wasn't wearing her apron.

"There's a phone call for you, Melina. I was just coming to say good night to you when I heard your name called. I'm on my way home now, but I'll see you again tomorrow." They were walking toward the phone setup. "How nice is that? You were dying to use the phone and now someone is calling you." The apron lady was sharing the joy she saw in Melina's face. They said a hurried good night and Melina picked up the waiting phone. It had to be Myron!

"Myron!" she said into the phone, her voice trembling with the need to hear his voice.

"Its me, Melina.. Roseanna – Myron's sister." Roseanna's voice sounded weak, faraway. Melina started to tell her sister-in-law all that was happening. Roseanna interrupted her. "I know, Melina. It is all over the news. I even know where you are. The airlines have confirmed it."

"Where is Myron? He must be home by now. Have you heard from him?"

"Melina, I have terrible news. I don't know how to tell you this. I wish there was another way." Roseanna's voice shook. Melina could hear her sniffling. "Myron was on the plane that went down in Pennsylvania. There were no survivors." The last words came over the long miles louder than the others, as if Roseanna was glad to be rid of her crushing news.

"You're wrong, Roseanna. They are saying that plane was headed for Washington. Myron wasn't going to Washington. He was flying to San Francisco! He –"

Roseanna interrupted. "Melina, listen, please! Flight 93 left Newark. It was headed for San Francisco. The hijackers took over the plane and turned it around. The news is saying some of the passengers tried to stop them, and that's why it crashed into the ground and not the White House."

Roseanna went on, telling her sister-in-law all she knew. How there were phone calls from Flight 93 before it went down. It was being reported that a man's voice from one phone call was heard shouting "Lets roll" just minutes before the airliner crashed. Roseanna was almost shouting into the phone, relieving her own grief by sharing it with another. But Melina wasn't listening. She just let the phone fall from her trembling hand. The receiver swung like an off-balance pendulum against walls that were closing in around her.

* * * * *

OUTSIDE, NIGHT WAS COMING to this northern land that was now so much an unwanted part of Melina's life. The sun had made its westing. An unseen truck whined on the nearby highway, following the fading day through the darkening boreal forest. Men and women walked around the filled parking lot. Some of them smoked, their cigarettes glowing brightly before fading away. Their talk sounded low and pleasant. Someone laughed, a pleasing, sharing sound of humans gathered at evening time.

Standing alone and leaning against a shadowed corner, a

slim figure also watched the earth magically transform itself from bright day to majestic night. Instinctively Melina turned south toward where her son had lost his life. Rising bright and full just above the distant black ridge, a bright star shone. It didn't twinkle like the others that came after, but kept its brilliance steady. It was Venus, that first of all the evening stars, that would still be there and shining last among the fading morning stars. Melina stared into the thrall of the beckoning star the ancient Romans had called the Goddess of Love. Funny, she thought, how stars just seemed to appear without actually seeing them come. *The sky is black, and then like magic they are there.* She couldn't believe she was even thinking about stars, when her world had once again fallen into turmoil.

The night chill came and tickled the length of her sorrowing body. She looked away from the silky firmament. She suddenly wished her apron friend were near. Melina gathered herself and walked away, turning her back on the last night she would see on this northern isle.

The first one at the travel agency next morning was Melina. The pretty, well-dressed young woman behind the counter prepared to launch her usual well-rehearsed speech for another of the plane people before Melina cut her off.

"Can you arrange an overland passage for one, please, to San Francisco, California, or anywhere in the States?" Melina's voice was clear, precise, polite, and very determined.

Melina realized for the first time since she arrived here that she really was on an island. The only way out of here was to fly, or by boat – a large car- and truck-carrying ferry, but a boat just the same. A person couldn't just drive away. It didn't matter to Melina. Her mind had been made up during the sleepless hours of last night. She would never fly again. Flight had taken everything she held dear away, never to return.

It took some time, but arrangements were made for her to travel by bus across the island and then by ferry to mainland Canada, and finally rail – first-class – overland, all the way to San Francisco. She was to leave by bus for the first leg of her long journey that very morning. The travel agency people were very efficient.

Melina got a taxi back to the Silent Witness Memorial. She knew she would never see it again. Standing beneath the tall, bronze soldier's image, she fancied she could see a resemblance to Michael in the sad, caring face. *Everyone sees what they want to see,* she reasoned. She turned her back on the metal figures and sat on the same bench as before. There was no one else around. The taxi driver stayed downhill, sensing her need for privacy. The events of her life came easily and unbidden. Strangely enough, they all seemed to centre around this small town. It was here that Michael had died. It was where his spirit rested. She knew that. Here, where she had found out about Myron.

Melina stood and walked back to the silent witnesses. She noticed a thin streak of patina separating from the boy's left eye. She hadn't noticed it before. She reached out slowly, hesitating yet wanting to touch the boy's fingers, hating the cold contact she had felt before. Closing her eyes, she let her fingertips brush the bronze hand. It was warm! She pulled her fingers away in surprise, then reached out and grasped the bronze hand in both of hers. Melina was sobbing again, but it was a cry of relief, an emotional rush she would never be able to explain. She had her farewell.

Melina never saw her apron-clad friend again. She wasn't

there when Melina collected her things for the bus ride across the island. Driving through the town of Gander, the bus passed a residential area. The day was warm and pleasant. Children were playing in well-groomed yards. A dog barked as the bus passed. And then Melina saw it. Surely there couldn't be two exactly the same. There, hanging from a high clothesline and flashing its colours gaily in the autumn breeze, was the wonderful apron she would remember for the rest of her life, waving at her. Melina waved back.

ACKNOWLEDGEMENTS

I would like to thank the following for their invaluable input toward the creation of this book: Glenn Blandford; Tom Bragg; Dave Brett, current fire chief in Gander; Shelley Brady Capota; Louis Collins; Theophilus Collins; Maurice Geange; Cynthia Goodyear; Pat Kane; Walter Kaplin; Sandra Kelly; Dr. Peter Matthews; Pam Stoucy and Michelle Stuckless of the Gander Public Library; Bill Ziegler.

Constant thanks to my wife, Rose, for all her invaluable help with my work. She is still my wife, my lover, and my very best friend.

Special thanks to Robyn Stack and Mary Froelich, who shared so much pain with a stranger. Also, I would like to express my gratitude to the administrators of the website "Gander: The Untold Story."

Thanks to Flanker Press for their continued confidence and encouragement in my work: to Jerry Cranford, my personal editor, who brings it all together so very well; to Margo Cranford, any author's best promoter, I would not be on as many shelves without her.

APPENDIX I

THE *POLLUX* AND *TRUXTUN*

THEY CAME UP OVER the rim of a black ocean horizon, three shrouded leviathans hidden in a moonless night. The USS *Wilkes*, flagship of the small convoy, was a destroyer, as was the older, four-stacked USS *Truxtun*. The supply ship USS *Pollux* completed the United States navy trio and, under the direction of their flagship, was leading the way into a night of North Atlantic hell.

It was February 18, 1942. Their nation had been at war for barely two months. The sneak attack by the Japanese nation on the American naval fleet at Pearl Harbor, on a clear Sunday morning, December 7, 1941 – all without a declaration of war – and the immediate response of their commander-in-chief, Franklin Delano Roosevelt, against the nation of Japan, now made it a true World War.

Before this brutal half-decade of conflict would end, no less than fifty-five million women, men, and children would die.

Before this night's voyage was done, some of the Americans aboard these ships would have their own names etched on that casualty list.

Even the day before, on the seventeenth, when the three vessels had cleared Cape Sable on Nova Scotia's southernmost coast, there had been no sun to aid the navigators in their calculations. A sparse showing of stars at the early part of that dismal night to "shoot," in order to confirm his bearings, had given the navigator of the *Wilkes* incomplete angles, which would prove to be devastating.

He took a brief fix on the stars Vega, Mizar, Dudhe, Antares, and Arcturas. The star Antares, brightest star of the constellation Scorpius, wasn't bright at all that night. In fact, it was barely visible. Antares was rising at the time of the scant sighting and appeared to be east of the meridian. The navigator, out on the cold, wet, rolling bridge of the *Wilkes*, calculated it to be west of the meridian. Under different conditions, with lots of sea room, this error would be minor; but they were headed for a narrow bay fraught with reefs, in a winter gale with a sideways-driving snowstorm.

They were following a "dead reckoning" course used by seamen the world over. This is the name given to a course taken from a ship's log and compass when the sky's celestial bodies could not be seen. It is never one hundred per cent accurate.

They had left Portland, Maine, and sailed in formation northeast toward and along the length of Nova Scotia. A raw, stiff north wind bore steadily around and came out of the southeast. It was a tearing gale that pressed heavily against the starboard sides of the warships as they plunged across the open mouth of the St. Lawrence gulf.

Long before the dark evening had plunged into a pitch-black night, they were ordered to follow the zigzag pattern. This wartime method of travel had been devised by the British admiralty as a way to confuse German submarine commanders who harried their prey all over the four million square miles of North Atlantic ocean. Convoys of ships, sometimes numbering in the dozens, would leave the usual ports of departure in St. John's or Halifax following a meeting between all captains and their navigators, to discuss the course to be followed during the treacherous crossing. They almost always used the zigzag pattern. It was a base course – true – laid down for the entire convoy.

At predetermined times, by order of its commanding officer, each ship would leave this base line and "zig" away on an abrupt angle to port and return and cross over the same line to "zag" away to starboard at another set angle. On paper it seemed straightforward, but at sea, with thirty or more ships involved, all of them capable of sailing at different speeds, it was far from simple. The speed ordered, according to the slowest ship in the convoy, was not always followed. Add to this a darkened bridge, blustery weather with high seas, radio silence between ships, and it was a navigator's nightmare.

It was this method of sailing that the three American ships were following on this night. Their preset time to begin zigging and zagging was every ten minutes. They had put Cape Sable behind them and were on a course for Cape Race, Newfoundland. Their base line – true – was 047°, a direct course set for the American naval base in Argentia, Newfoundland.

Chosen in 1940 to be an American naval base under the U.S.–British Lend-Lease Agreement, Argentia would be occupied by U.S. Marines as early as January 25 of 1941. On February 13 of the same year, the U.S. flag was raised, making it all official. By the summer of 1941, Argentia was recognized in every Allied command. It was the nearest American naval base on the North American continent to the war raging in Europe. The Stars and Stripes would fly over Argentia for the duration of World War II.

Argentia's importance in the war effort was further sanctioned on August 10, 1941, when Winston Churchill, prime minister of Britain, and President Roosevelt of the United States met aboard ship in Argentia's deep harbour – locally called Little Placentia Bay – and signed the Atlantic Charter. The safe harbour the seventeenth-century French had called "Petite Plaisance," or "pleasant little place," had become a depot of war. It was this naval and home base the three American ships were sailing toward on this winter's night.

Now each ship that left the true course rolled like a drunken sailor who veers suddenly from his road home and staggers away into the darkness, only to reappear farther along the same deserted road and stumble the rest of the way home. Of the three ships, the *Truxtun* probably fared worse in the deteriorating sea conditions. The wind was building rapidly from the southeast and would soon reach force seven, a bad blow. It brought with it a driving snow that plastered the starboard sides of the ships and started a buildup of ice on their superstructures. The *Truxtun* had a beam of only thirty-one feet, her 310-foot length keeping her in a constant wrenching roll that sickened her younger sailors and frightened the older. At times the *Truxtun* rolled so badly the smoke appeared to drain horizontally out of her four ancient steel stacks.

The *Pollux* carried sixteen crewmen for the USS *Prairie*, which was already at the Argentia port. The *Prairie* had hosted Task Force 24 of the United States Atlantic Fleet in the summer of 1941 in Argentia and, under Vice Admiral Arthur LeRoy Bristol, established flag headquarters there. The *Pollux* also had fifty-eight raw recruits headed for training in Argentia. They had never experienced such a night as this. Metal buckets fastened along the insides of all of the ships stank of vomit.

The Cabot Strait, famous for its powerful tides and funnelling winds that bear down from the maw of the Gulf of St. Lawrence, lived up to its reputation on this night. It would get even worse as the trio entered the waters off Newfoundland's south coast. The drift of current here almost always moves toward the land. It is a powerful flow of tide barely under the surface, backed by a strong winter gale from the southeast that further urges the sea to the land, especially at night. Seamen who dare the treacherous channel are required to be more than vigilant.

All three ships were equipped with a light-lock system that prevented any glare to pierce the dark and potentially alert the watching eyes of any prowling German submarines. When the door latch was released on either the port or starboard sides of the bridge, a switch was activated and even the faint lights from inside were automatically turned off. They could only be turned back on again from the inside.

Another great worry for the captains of the three ships was collision with one of the other vessels. They ordered the watches to keep a constant lookout, but this was extremely difficult. Great skeins of stinging wind and snow lashed at their eyes as they peered into the tormented darkness. Monstrous waves met the puny manmade floats upon which the watchmen stood and, finding them in their way, flung their fierce defiance up and over the plunging ships. Spray soaked through every seam of the men's clothing. The ocean roared its might and deadened their senses with a numbing, never-ending cold. As the night wore on, the navigators, unable to get an accurate reading from the sky, put their trust in the compass bearings and the ocean bottom below their submerged keels.

The fishing banks around Newfoundland were known the world over. Their generous grounds were home to one of the world's richest supplies of protein. These rich fishing banks were also well documented on ships' charts. The fathometers aboard all three ships gave them a steady bottom reading. While this was not one hundred per cent accurate for absolute positioning, it was an excellent indicator of their line of travel.

On board the flagship *Wilkes*, a young seaman recorded the fathometer readings from the St. Pierre bank as the great ship passed over. St. Pierre was in their path, all right, but the problem the inexperienced sailor failed to note was the time. They were supposed to cross the St. Pierre bank at 0138 hours, but they were actually passing over the well-known fishing bank at 0023 hours. They were one hour and fifteen minutes ahead of schedule.

The convoy was dangerously off course, due in part to the terrible gale assaulting their starboard bows, and also the south-flowing current. The three ships were perilously close to an impassive landfall. The navigator of the *Pollux* calculated that they were indeed off course and suggested to his commander a huge 30° change. They were too far to the north, he insisted. After some arguing the course was altered by only 10° – not nearly enough!

The change of course was blinked to the other two ships by a small, six-inch beam of light. Even on such a night as this, submarines were a constant worry. The fury of that terrible storm didn't allow much feeble light penetration. By now visibility was almost zero, and the other ships did not read the change in course heading.

The *Wilkes* struck land first. Only minutes before, her fathometer had shown thirty-five fathoms, then fifteen fathoms as the bottom came up quickly to meet the unsuspecting ship. By this time the high, white land loomed out of the blizzard. The lookout thought it to be a huge iceberg, but it was the stern, forbidding coast of the Burin Peninsula on the south coast of Newfoundland.

* * * * *

FIRST CAME THE SHOUTED order, "Stand by for collision!" Then, "All engines astern, full emergency!"

But it was too late. The *Wilkes* was aground. Her searchlight flashed through the fury of the blinding snow to warn the other two ships, but her warning came too late.

Within minutes the *Pollux* and the *Truxtun* had also slammed into the mighty "boot" of Newfoundland's south coast. The unthinkable had happened, but it would only get worse.

The *Wilkes* proved to be only grounded forward. After discharging all possible bow weight – including her two steel anchors and thirty-eight fathoms of steel chain – and though her bow compartments were flooded, after many tries she managed to pull herself free. She headed back to the relative safety of the open sea. The others were not as lucky.

Closer to the land, below a headland known as Lawn Point, a crewman had seen the beacon flashing from the *Wilkes*. At the same time he also saw the snowbound land rushing to meet them. From his frozen throat, as well as from the masthead lookout and both starboard and port watches, came the same terrifying yell: "Land dead ahead!" The frantic cry was of course too late. The *Pollux*'s engines rammed the 15,000-ton ship full into the sheer cliff, which split her port side like a can opener.

The lookout aboard the *Truxtun* had also seen the *Wilkes*'s warning light too late. A reported "light spot in the sky" was the only way of describing what they saw. It could have been a scud of snow-laden wind, for all anyone knew. By now the storm was in full fury. Before the meaning of the faint light became clear, the terrible land appeared through the squalls no more than 200 yards to port. The starboard engines backed full astern. Her propeller was torn away. The ship swung 10° or so to the right, enough to take her away from the serrated cliff of Lawn Point, but only to wedge the doomed vessel onto the rocky bottom of Chambers Cove. The seas pressing against her stern lifted her up and drove her mercilessly toward the land. She would not leave this cove again.

Waves higher than the ships themselves rolled in from the angry Atlantic and pounded them without relent. Now began a night of terror.

Aboard the two ships, orders were shouted at men who clung to tilted, swaying, wrenching, sea-flushed decks. The only sign of light in that interminable night was the fuming white waves that dashed endlessly down upon the doomed vessels.

Masters of ships and men were now faced with an ordeal they were ill prepared for. These men, who were capable of dealing with the most formidable situations at sea, with lots of room between land and keel, were suddenly thrust into a new and unimaginable environment. They were neither at sea nor upon the land, but caught in a terrible void where land and sea collide. Masters and their charges would be taxed to their extreme limit. Many would be found wanting.

The night raged around them and time was not on their side. The ebb tide would soon turn to flood, bulging the taut North Atlantic muscle.

* * * * *

FROM THE SHIPS' STORES they brought pieces of planking, which they nailed end to end in a vain effort to reach shore. Rafts proved useless. Lifeboats were crushed against the steel sides of the vessels before they could reach the roiling sea. More became welded to their steel davits with ice. One finally made it to the strand using a line from one of the vessels. Its soaked and frozen occupants staggered up the last few feet to escape the sucking sea. A raft finally made it.

The seamen fastened a bigger, stronger line to the two ships. They rigged a breeches buoy, and finally rescue was begun. Men jumped into the sea. Some of them fell. Many of their bodies were squashed as easily as one of the frothy spindrift of bubbles that laced the frenzied cove.

Getting through the freezing, oily water and finally standing on shaking legs was only half the battle facing the terrified sailors. Now they must somehow scale the icy cliffs, whose heights were hidden beneath the swirling sky.

Exhausted men hawked black, salty mucus out of their

chattering jaws. All who had been in the water were plastered with the glutinous crude that was spewing out of the ruptured tanks of the ships. Some of them were like white actors painted as black men in an old Vaudeville act, their white teeth and bulging white eyeballs glittering in the dark.

The tide had turned and, along with the wind, was now reaching farther inland. The distance between the sea edge and formidable heights was shortening. A young sailor decided to try and climb out of the desperate prison. He threw his muscled body against the frozen land and, using his clasp knife, began carving handholds. An older sailor followed him. The two men climbed up over that mountain of misery, their struggling limbs warming their bodies for the task. No sound escaped their lips, but from below came a thin chorus of encouragement as the two men hung on like desperate rats, finally disappearing over the swirling ledge above.

Whatever the two men who had escaped the clutches of the cove had expected to find, they didn't. There was only an endless, crashing wash of white coastline to their right, and a dense, undulating forest stretching away to their left. The headland sloped downward ahead of them. They attempted to walk down, but both men lost their footing and slid like schoolboys down over the icy slope, dark streaks of oil trailing along behind them.

When they reached the stunted trees at the bottom, they found themselves out of the deafening noise and raw bite of the wind. It was a relief to the two men, who were numbed from head to toe. The snow had quickly frozen into a chilly blanket on their wet clothes.

They stumbled farther down the ravine following the treeline. Amazingly, in that dark night, they came upon a small wooden shack. It was also deserted. The door was unlocked. Inside they found dried hay. This singular mark of human habitation gave them courage; someone must be living nearby. They considered their options, then quickly decided that the younger of the two would go to find help. The older man buried himself in the hay in an effort to keep warm. Soon other survivors from the wreck sites stumbled into the shed.

Away from the shack and the sheltered gulch, the young sailor made his way. To his right the seas rolled toward the land in such a fascinating display of might that, in spite of his misery, he was awed by the sight. Where the water reached up as far as it could over the land, high sheens of spray borne by the unrelenting wind travelled farther still. The droplet remnants of the waves fell upon the white land and down upon the young sailor in frozen ice pellets. He kept away from the bluff and headed through the scrub forest.

He was borne along by a strong will to live, despite the worry that the sea was taking a deadly toll on his shipmates still below in that cove of terror. He would press on until he could go no farther. With the blood coursing through his body keeping him in sedate numbness, he no longer felt cold. His calf muscles burned with each step as he moved through knee-deep snow. His breathing came in quick, laboured gasps through clenched teeth. He had no idea where he was.

The young seaman couldn't believe that a wood could be so difficult to get through. There was no pattern to the wooded growth, nor any discernible space between the trees; each one intertwined with its neighbour. They were short, just a little over his head. He was in a jungle of small, twisted trees that was almost impossible to get through. Once he thought he crossed a narrow trail, but in the darkness he lost it again. He waded across a narrow, swollen brook, the rushing water making his rough trousers cling tightly to his skin. On he stumbled.

A thin, grey daylight came. The trail ahead grew more distinct, but he barely noticed in his exhausted state. He struggled up over another ridge and, staring into the brightening distance, he saw buildings. He stumbled on, surprised that he couldn't get his legs to go faster. He called out for help.

A man close to one of the structures turned his way, a puzzled look on his face as the hatless, uniformed man caked with ice walked toward him. The sailor stumbled the last few steps and, leaning on the shoulder of the man who would begin the rescue, blurted out his pitiful story. The rescue had begun.

* * * * *

THE THREE SHIPS HAD run aground in a small indentation on Newfoundland's south coast locally called Chambers Cove. The USS *Wilkes* was the lucky one. It had managed to back away from the reef that had temporally grounded her. The USS *Pollux* and the USS *Truxtun* were doomed on this "heel" of the huge Burin Peninsula boot, which on a map appears to be light-stepping into the broad Atlantic Ocean.

The men from the town of St. Lawrence, which the oil-stained seaman had miraculously found, as well as the nearby town of Lawn, knew from experience no deliverance from the sea was possible in Chambers Cove.

Men on snowshoes and with horses and sleds and ropes

floundered back through piled snowdrifts. The young American sailor refused to stay behind, but followed back over his stumbling tracks in the blowing snow. When they reached the high edge of Chambers Cove in the full, windswept light of day, they were awed by the carnage below them.

Two once mighty ships, one of them fully accoutred for war, its long, deadly gun barrels pitching and heaving uselessly, were being battered without mercy. Less than a mile offshore, another grey vessel resembling the gunship prowled restlessly back and forth, powerless to assist. At intervals above the sound of breaking sea and roaring wind came the sounds of men screaming. The two ships were half seas under. Men were hanging on to their steel rails. Men were climbing down the grey hulls on swaying cargo nets. Men were clinging to bobbing lifeboats. Men were dying. Men were dead.

The thousands of gallons of oil spilling from the vessels with every sweep of the rollers was both a blessing and a curse. Its weighted, stinking scum calmed the roiling sea surface, but for men trying to keep their heads above water in a turbulent sea, it also meant death. The oil added a leaden weight to saturated clothing. It also kept precious air from gasping lungs when it filled open mouths. Oxygen-starved seamen went into involuntary bouts of coughing in a vain effort to cleanse their airways of the viscous fluid. Their eyes stung with the foreign mix of salt and crude. Their arms flayed with panic. Many of them sank below the surface. Some of them stayed there. The winds raged on.

Both the towns of St. Lawrence and Lawn began a rescue effort amid this chaos. St. Lawrence was closer to Chambers Cove. All that day and into the next, without cease the people responded as Newfoundlanders had always done. Many of the citizens were hard-nosed miners who had grown accustomed to hazards on a daily basis in the underground fluorspar mine near the town of St. Lawrence, the largest mining operation of its kind in North America.

Fluorspar is a non-metallic ore used in the manufacture of aluminum and glass. It is also used in the production of freon, the main ingredient for refrigeration. Few safety measures were used in the extraction of this ore from the subterranean depths just beyond their doorsteps. Just the year before the miners had asked for nothing less than a government tribunal to have the hard-working miners undergo a medical examination. So far their request had fallen on deaf ears. These men of the deep were sucking down the deadly "silica dust" daily. Without even knowing it, many of them would later succumb to suffocation from silicosis.

Others from the area were fishermen who knew the sea at their doorstep so well that a mere glance from their windows could tell them what each day would bring. There would be no relief from the winds on this day. Not only that, they also knew that, even when the winds abated, it would take days for the heavy swells the storm had brought to diminish. There would be no small boats going into Chambers Cove any time soon. Rescue would have to come from the land.

For the next few days these simple, hard-working people would bare their homes and hearts to desperate strangers. Men made countless trips out to Chambers Cove. They scaled icy cliffs and plunged into freezing waters. They pulled men drunk with cold and fatigue from the sea. They hauled them up over the icy capes to safety. They carried them on their backs. They waded into the oily surf and pulled dead bodies too late to the saving land.

* * * * *

WOMEN WHO ALWAYS STOOD by their men were as much a part of the rescue as the men themselves. Oil-soaked, near-dead sailors were laid on spotless featherbeds that would never come clean. Warm couches by ever-burning stoves became comfortable beds for many. Soup and teakettles were kept simmering day and night. No one would want for the nourishment and comfort the task out at Chambers Cove demanded. Wives and daughters kept the "home fires" burning for the men and boys who by now had beaten a well-worn trail through the snow out to the wreck site.

Inside these well-kept homes trudged snow-covered men bearing oil-coated burdens, all in need of immediate attention. Some had broken bones. Others had cuts and severe bruises. The debris tossed around in that wild cove had shown no mercy to the frail humans trying to escape the rumbling sea. Almost all of the seamen were coated with oil. It stank up every kitchen they entered. Women set to work with hot, soapy water and their best cloths to clean them.

One of the smelly, blackened men received more attention than the others. Still, no matter how many times he was scrubbed, his skin remained black. It was only when the man regained consciousness that the mystery was solved. He was a genuine black man. He was the first true African American any of the villagers had ever seen. The respect and simple human compassion he had received from these people would change the man's life forever. For the first time in his life, he was treated as an equal. With respect, with a kindness he had never before received from white people. For the first time in his life he was just one of the boys. This man, Lanier Phillips, would never forget it.

Several of the women nursing the American seamen back to health still had a vivid memory of another disaster. Just thirteen years before, no less than fifty small outport communities had been dealt a deadly blow, betrayed once again by the very ocean that sustained them. Monday evening, November 18, 1929, an earthquake far out to sea – no more than a tremor here on the south coast of Newfoundland – had unleashed a terror never before seen in this corner of the Atlantic.

Shortly after the tremor had gone away, the seas calmed. It was an unnatural calm, the ocean so still as to be devoid of all life. The water receded out of the bays, farther out than any normal ebb tide had ever taken it. Everyone stared toward the calm, silent ocean. Everyone waited, not knowing why. Then the answer to the strange rumblings and earth movements was revealed.

Out beyond the indented bays and past the high headlands, a grey wall of water suddenly appeared. It seemed to be endless, reaching the length of the flat horizon. At first it came soundless, even when it poured its immense weight onto the flat surface, as it raced shoreward. Then came a dull roar of sound. It sounded like a faraway roll of thunder on a summer's day. Each time the wave broke over itself, it seemed to feast upon the sea, for it rose mightier than before, attaining even greater heights until flinging itself forward again, its cresting roar finally reached the land.

Two more waves followed. Ironically, Newfoundland fishermen say big seas always come in threes. When the last one finally retreated from the battered coast, the tsunami had damaged the length of the south coast and had laid waste to the lives and livelihoods of entire towns. Boats and fishing rooms were splintered and tossed far inshore by the great thrust of the waves, or sucked back out to sea when the waves flattened again.

Nor were the homes of fishermen spared. They were crushed and swept aside where they stood, or plucked from their foundations and swallowed, to be finally spat, pulverized, out of the great mouth of the sea. Some homes were even swept out fully intact, like bloated carrion birds, away from the sorrowing villages.

Twenty-seven lives succumbed to this onslaught from a merciless ocean. As many as 10,000 people suddenly found themselves without a roof over their heads. Lives were changed forever. Whole villages were devastated. In some cases the trust that men and women had for their environment was shaken. The island wept collective tears. The people of the south coast of Newfoundland stayed and mourned their losses and helped each other the way they had always done. They rebuilt their homes and lives and went back to sea again. It was as if the ocean would always have them in its thrall.

This was the stalwart race of people who had come to the rescue of the American sailors. The foreigners were in good hands. The people of the town of St. Lawrence, without giving it much thought, had kept alive the name of their patron saint. When the ruling members of the Roman Catholic Church had demanded that Lawrence, then a deacon, produce all of the wealth of his church, he gathered all of the poor and blind, the crippled, diseased, and orphaned. When he shouted, "These are the treasures of my church," he was martyred.

These people of the sea refused to give up their efforts to save the American seamen stranded and dying on their shore. Straining muscles pulled a dory overland to Chambers Cove, across the trampled snow trail, and lowered it by stranded hemp ropes down the cliff face. Into the churning sea they rowed, out to the wretched wrecks, to pluck crying sailors from the edge of death. Sadly, even the heroic efforts of hardened seamen would not be enough to save all. One hundred and eighty-six of the men from the *Pollux* and the *Truxtun* survived the sinking of their ships. Men and women who faced tragedy from the sea as second nature had never seen anything like this. Despite the constant struggle of locals, as well as survivors of the wrecks themselves, the sea called its deadly roll. Two hundred and three of the Americans did not answer. Thanks to the heroic and selfless efforts of the people of the south coast of Newfoundland, 186 of them did.

It was one of the greatest death tolls of American servicemen ever on friendly soil. Sadly, it would not be the last for the Americans on this island.

APPENDIX II

Screaming Eagles

We have a rendezvous with destiny.
Our strength and courage strike the spark,
That will always make men free.
Jump right down through the skies of blue,
Keep your eye on the job to be done.
We're the men of the hundred first,
We'll fight till the battle is won.

The above song was performed at the 3/502 Division Memorial ceremony in Fort Campbell, Kentucky, on December 20, 1985. President Ronald Reagan referred to this song in his address on that day.

The following scripture verse was read as well:

Isaiah 40:28-31

Do you not know?
Have you not heard?
The LORD is the everlasting God,
The Creator of the ends of the earth.
He will not grow tired or weary,
And His understanding no one can fathom.

He gives strength to the weary
And increases the power of the weak.

Even youths grow tired and weary,
And young men stumble and fall;

But those who hope in the LORD
Will renew their strength.
They will soar on wings like eagles;
They will run and not grow weary,
They will walk and not be faint.

The 101st Airborne Song – Screaming Eagles written by Captain Samuel R. Loboda

GO! You Screaming Eagles
Dive thru the sky!
Never let your strength and courage fail.

Just make your jumps
And feel your parachutes
Grab the air with a snap
As down to earth you sail.
Come on let's GO!
You Screaming Eagles
Fight for the right to be called a son
Of the "ONE-O-ONE"
So here we GO!
You know the Eagles are
Diving from the sun,
Look out for the MIGHTIEST MEN of
Ge-ron-I-mo!

Darrin Patrick Brady's sister, Shelley, wrote the following poem for him.

TEARS

The caring
The sharing
The love
The anger
The hate
The nightmare
The fear
The prayer that never got answered
The hoping that he'd call
The shock

The hurt
The loss of a loved one
No hellos
No goodbyes
No I love yous
Just tragedy
The pain, will it ever go away?
And knowing.
That your big brother won't ever be coming home again.

APPENDIX III

JOHNNY ALLEN HENDRICKS, LATER renamed James Marshall, was born November 27, 1942, in Seattle, Washington. He died of a drug overdose in London, England, on September 8, 1970. He was buried in Greenwood Memorial Park in Renton, Washington.

Considered by many to be the greatest electric guitarist who ever lived, he was better known as Jimi Hendricks. He was a left-handed player. Hendricks had started his amazing, all-too-short career as a ukulele player. He was the first person to be inducted into the Native American Hall of Fame.

Jimi had reluctantly served some time in the American Armed Forces. After a run-in with the law, he was sentenced by a court judge to serve two years in the U.S. Army. He spent less than one year as a paratrooper in Fort Campbell, Kentucky, with the 101st Airborne Division. Discharged with a broken ankle, he did not return to the Forces.

APPENDIX IV

PHOTOS

WHERE EAGLES LIE FALLEN

IMAGE BY CLINT COLLINS

U.S. Serviceman Darrin Patrick "Buckeye" Brady

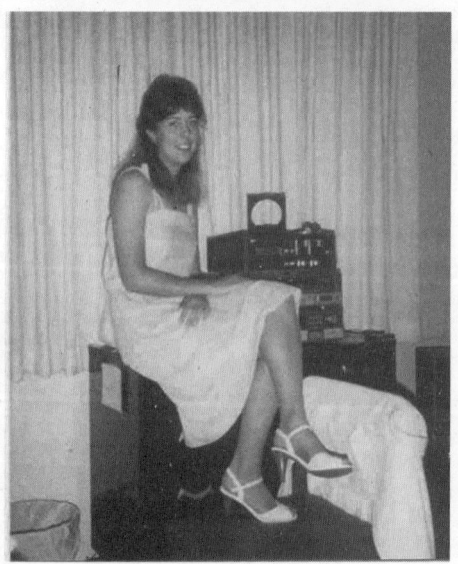

U.S. Servicewoman Cathy "Ziggy" Ziegler

Cathy Ziegler riding a camel in Egypt

Virginia Ruth "Jenny" Word (left) and Cathy Ziegler (second from right) with friends in Egypt

Sergeant Robert Stanley Kaplin in the Sinai Desert in 1985

Robert Stanley Kaplin and his mother, Joan, at his Confirmation

Robert Stanley Kaplin (left) and his best friend, Mark Decocq

1940 Mercury Coupe, named *Screaming Eagle*, built and owned by Walter and Joan Kaplin of Gig Harbor, WA. "Dedicated to 248 members of the 101st Airborne, the fabled 'Screaming Eagles' who perished in an airplane crash at Gander, Newfoundland, December 12, 1985. Our beloved son, Sgt. Robert S. Kaplin, age 24, also perished in the crash."

Formal dedication of the *Screaming Eagle* at the first Mild to Wild Car Show at the Tacoma Dome in November of 1989. The radical custom 1940 Mercury Coupe was built by Walter and Joan Kaplin, co-owners of the Mild to Wild Car Show from 1989 to 1991. Pictured are members of I Corps from Fort Lewis, Washington, and other retired military who were also part of the dedication ceremony.

Sandra Kelly today, who was deputy mayor of Gander, Newfoundland, in 1985

U.S. Serviceman Michael Shayne "Eliot" Stack

Wedding of Shayne Stack's sister Terry in 2005. L-R: Nancee (Musil) Foeller, Kaye Blew (Shayne's grandmother), Robyn Stack, Bill Stack, Fred Musil, Teresa (Stack) Dure, John Dure, Jacob Stack, Mitzi (Musil) Chervenka, Lauren Stack, Jeni Diprizio Stack. Immediate family missing from the photo are Scott Musil and Shayne's older sister, Stormy Stack, who died in 2003.

U.S. Serviceman Frank Wheeler Froelich

Mary Froelich, mother of Frank Wheeler Froelich

APPENDIX V

CASUALTIES OF ARROW AIR FLIGHT 1285
From the plaque erected by personnel of
9 Wing Canadian Forces Base Gander September 1995

PFC Abrams M.	PFC Council O.	PFC Hansen W.
Sp4 Alexander H.	Pvt 1 Craig M.	Pvt 1 Hardeman C.
SSgt Andrehoff S.	Sp4 Crawford P.	CWO3 Hardin B.
Sp4 Andrews D.	SSgt Cruzalgado F.	PFC Harris B.
Sp4 Aponte I.	Pvt 1 Cupples T.	1st Lt Hart R.
PFC Arrowood S.	Sp4 Daniels W.	Pvt 2 Hassing M.
Sp4 Arvin R.	Sp4 Danielsen T.	Sp4 Haugsdahl R.
1st Lt Avillan L.A.	Sgt Davis J.	PFC Heidecker D.
Sp4 Banks B.	Sgt Davis J.	Sgt Hemingway P.
Sp4 Barber D.	SSgt Davis T.	Pvt 2 Highfill J.
Sp4 Baumann E.	Sp4 Deckman H.	PFC Hileman T.
PFC Beer E.	Cpl Diventura J.	Sp4 Hobbs D.
Sp4 Benson W.B.	Sp4 Dixon T.	PFC Hobbs K.
Sp4 Bittle S.	SSgt Duckworth J.	SSgt Holliman J.
CSM Black H.	Sgt Dumpert B.	Sp4 Hoyer R.
Sp4 Bostwick P.	Sgt Easley B.	Sp4 Hughes C.W.
Sgt Bowen J.	Capt Eastman M.L.	SSgt Hughes F.
CWO3 Bowen R.	Capt Edmonds K.	PFC Hull J.
Sp4 Bradley J.	Sp4 Englebert C.	Cpl Ivy H.
Sp4 Bradshaw S.J.	Sgt Ferguson J.	Pvt 1 Jackson A.
PFC Brady D.	Sp4 Ferguson M.	Lt Col Jeffcoat M.
Sgt Brancato C.	PFC Fink K.	SSgt Jennings D.
Sp4 Brasfield T.	Sp4 Fitch D.	Sp4 Jennings T.
PFC Brilya W.	Cpl Foskey T.	Pvt 2 Johnson J.
Sp4 Britt G.	Cpl Fuller P.	SSgt Johnson R.L.
Sp4 Brown J.	Sgt Gantzer K.	Sgt Jones J.
PFC Buchanan G.	Sp4 Gayton A.	PFC Jordan D.
Cpl Burdette J.	Sp4 Gerdes S.	Sp4 Kaplin R.
PFC Bury D.	Sgt Givens G.	Sgt Karadsheh I.
PFC Campbell T.	SFC Godsey D.	Sp4 Kee J.S.
PFC Carter G.	Sp4 Gonzales M.	Sgt Kidd T.
Sgt Carter M.	PFC Gonzalez R.	Sp4 King J.
Capt Carter T.	Pvt 1 Gorree J.	Capt King R.
Pvt 2 Cartwright D.	PFC Graham K.	SSgt Kirby T.J.
PFC Caudill P.	Sp4 Graham T.	PFC Kiser B.E.
PFC Chaddock G.	SSgt Grala D.F.	1st Lt Kosh J.K.
Sp4 Colby S.	PFC Gray C.	SSgt Kubic M.R.
Sp4 Coleman B.	PFC Guerra R.	Sp4 Kuehn J.M.
Sgt Cordero M.	Capt Haller B.	Sp4 Lane R.A.

Maj Lawrence M.R.
Sgt Lineberry. D.G.
Sp4 Lloyd W.M.
1st Lt Long P.
Sp4 Lundgren D.C.
Sp4 Lynch B.R.
SSgt Malone J.W.
Capt Manion E.J.
Pvt 2 Martin T.L.
Pvt 2 Mathis D.L.
Sgt Mayhew R.G.
Sgt Mcardle P.A.
1st Lt McCarty J.
Sp4 McCleery C.M.
1st Sgt McCook R.F.
1st Lt McCormick S.J.
Sp4 McWhite C.
CWO2 Miller D.
Sgt Miller L.G.
SSgt Miller R.
Sp4 Miller T.E.
Sgt Millett J.M.
Sgt Mollett J.A.
SSgt Moore S.T.
PFC Morgan L.
Sp4 Mullins S.W.
Sgt Murry M.
Sp4 Napler M.A.
SFC Nartia J.A.
SFC Nelson D.C.
Pvt 2 Nelson K.J.
SSgt Nelson S.R.
Sgt Nichols R.S.
PFC Nolan M.T.
Sgt Ocasio F.
Pvt 1 Olson R.L.
Sp4 Owens G.A.
Sgt Padgett G.W.
PFC Pafford T.L.
Sgt Palmisano J.R.
CWO3 Parris R.
Sgt Parsons T.
Pvt 2 Perry V.S.
Sp4 Pevey T.R.
Pvt 1 Phillips A.
Sgt Phillips J.
1st Lt Powell B.C.
Sp4 Puntanen R.K.
Sp4 Rahr M.R.

Capt Rains T.L.
Sp4 Rawls D.L.
Sp4 Reasbeck P.
Pvt 1 Reed M.W.
SSgt Reynolds J.T.
PFC Richardson G.
PFC Rimiller R.D.
Sp4 Roberts B.E.
Sgt Roberts W.D.
Sgt Robertson V.
Sgt Robinson T. Jr.
Sp4 Russell R.C.
Sp4 Ruth R.A.
Sp4 Schmoyer R.
Sgt Schremp P.E.
PFC Schultz K.M.
Sp4 Scott G.L.
Sp4 Searcy B.
SSgt Sears R.W.
SSgt Sellner T.
PFC Seitz F.O.
PFC Serna E.W.
Sp4 Shipley M.
PFC Shook J.E.
PFC Shultz R.D.
Sp4 Simmons C.N.
Sp4 Simmons G.
Sgt Singleton E.
Sp4 Sloan M.S.
Pvt 2 Smith C.D.
SFC Smith R.V.
Sp4 Smith S.J.
Sp4 Smith T.V.
PFC Spearman M.
Sp4 Spears J.M.
Pvt 1 Stack M.S.
PFC Staten D.C.
PFC Stearn A.W.
Sp4 Stephens D.
2nd Lt Stevens K.L.
Sgt Stewart R.S.
CWO2 Stone E.
Sp4 Straub G.L.
Sp4 Stringer R.
Pvt 2 Stritch S.A.
Sp4 Thomas R.F.
Sgt Thomas R.K.
Sgt Thompson D.C.
Sp4 Thompson S.B.

Sgt Thornton C.G.
Sgt Travis T.
Pvt 2 Tucker T.N.
SSgt Turner V.L.
Sp4 Venneri S.C.
PFC Vinson W.
Sp4 Walker G.
Sp4 Walker G.W.
Sp4 Wallace B.E.
Pvt 2 Wallace M.E.
SFC Ward A.
SFC West T.E.
PFC Wester J.C.
Sp4 Wheeler F.C.
SSgt White E.S.
Sp4 Whiteman M.L.
SSgt Wilburn D.
Sp4 Williamson J.H.
Sgt Willingham R.N.
PFC Wilkins F.
Sp4 Wilson R.
SSgt Winston J.A.
Sp4 Wisson T.
2nd Lt Witmer J.B.
Sgt Witt K.M.
Sgt Wolford R.N.
SSgt Wood L.A.
Sgt Wooliver W.
Sp4 Word V.R.
Sp4 Wright J.R.
Sp4 Wyn R.
Sp4 Yeargan G.T.
Sp4 Ziegler C.M.

F/O Connelly J.R. (crew)
F/A Cutler S. (crew)
F/Eng Fowler M. (crew)
Capt Griffin J. (crew)
F/A Matasovski M. (crew)
F/A McKay D. (crew)
F/A Phillips D. (crew)
F/A Serafin J. (crew)

RESOURCES

Badcock, Captain T. C. *A Broken Arrow*. St. John's: Clouston Publications, 1988.

Brown, Cassie. *Standing Into Danger*. St. John's: Flanker Press, 1999.

Callahan, William R. *The Banting Enigma: The Assassination of Sir Frederick Banting*. St. John's: Flanker Press, 2005.

Cranford, Garry. *Tidal Wave: A List of Victims and Survivors, Newfoundland, 1929*. St. John's: Flanker Press, 1999.

Faulks, Sebastian. *Birdsong*. Vintage, 1997.

Filotas, Les. *Improbable Cause*. Seal Books, 1991.

Goff, Roderick B. *Crossroads of the World: Recollections from an Airport Town*. St. John's: Flanker Press, 2005.

Reader's Digest Editors. *Great Events of the 20th Century*. Reader's Digest Association, Incorporated, 1981.

Turnbull, Ann, and Joseph Wase. *Blows of Circumstance*. White Mane Publishing Company, 1989.

Other Sources:

Associated Press, December 14, 1985

Baltimore Sun newspaper, December 14, 1985

Burlington County Times newspaper, December 15, 1985

Evening Telegram newspaper, December 12, 1985

Evening Telegram newspaper, December 14, 1985

Evening Telegram newspaper, December 16, 1985

Evening Telegram newspaper, December 17, 1985

Evening Telegram newspaper, December 21, 1985

Gander Beacon newspaper, December 12, 1985

Gander Public Library

Philadelphia Inquirer newspaper, December 15, 1985

Films Related to the Screaming Eagles:

A Bridge Too Far. 1977. Directed by Richard Attenborough.

Band of Brothers. TV miniseries, 2001.

Battleground. 1949. Directed by William A. Wellman.

The Dirty Dozen. 1967. Directed by Robert Aldrich.

Hamburger Hill. 1987. Directed by John Irvin.

The Hurricane. 1999. Directed by Norman Jewison.

The Longest Day. 1962. Directed by Ken Annakin, Andrew Marton, Bernhard Wicki, and Darryl F. Zanuck.

Saving Private Ryan. 1998. Directed by Steven Spielberg.

The West Wing. TV series, 1999-2006.

GARY COLLINS was born in a small, two-storey house by the sea in the town of Hare Bay, Bonavista North. He finished school at Brown Memorial High in the same town. He spent forty years in the logging and sawmilling business with his father, Theophilus, and son Clint. Gary was once Newfoundland's youngest fisheries guardian. He managed log drives down spring rivers for years, spent seven seasons driving tractor-trailers over ice roads and the Beaufort Sea of Canada's Western Arctic, and has been involved in the crab, lobster, and cod commercial fisheries.

His writing career began when he was asked to write eulogies for deceased friends and family. He spent a full summer employed as a prospector before he wrote *Soulis Joe's Lost Mine*; he liked the work so much, he went back to school to earn his prospecting certificate. A critically acclaimed author, other books to his credit include *Cabot Island*, *The Last Farewell*, *Soulis Joe's Lost Mine*, and the children's illustrated book *What Colour is the Ocean?*, which he co-wrote with his granddaughter, Maggie Rose Parsons. The latter won an Atlantic Book Award: The Lillian Sheppard Memorial Award for Excellence in Illustration.

Today, Gary is known all over Newfoundland and Labrador as "The Story Man." His favourite pastimes are reading and writing, and playing guitar at his log cabin. He lives in Hare Bay, Newfoundland, with his wife, the former Rose Gill. They have three children and three grandchildren.

Gary Collins can be reached by email at **nicholasc68@live.ca**.
The official Gary Collins website is **www.garycollins.ca**.

ALSO BY GARY COLLINS

"Collins's focus on an ordinary event taking place under extraordinary circumstances sheds a tender, respectful light on how strength of character can be forged at the anguished intersection of isolation and bereavement."
Downhome

In November of 1954, a terrible storm darkened the skies above Cabot Island and battered its solitary lighthouse with a single-minded fury. The keepers of the Cabot Island light were no strangers to sea weather, but when tragedy struck the brothers Gill, the younger of the two was left to fend for himself amid one of the worst storms in Newfoundland's history. This is a true story of the love between two brothers, a love that perseveres in the face of death, loss, and greatest personal challenge.

ALSO BY GARY COLLINS

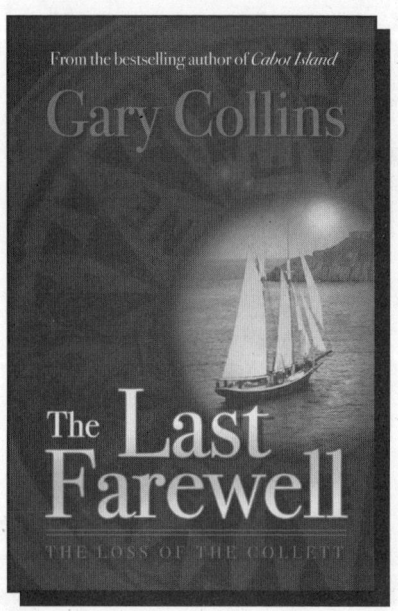

"Read *The Last Farewell* not only because it is a moving historical tale of needless tragedy but also because it's a book enriched with abundant details of Newfoundland life not so widespread anymore."
The Pilot

The Last Farewell tells the true story of a crew of logger-sailors who left their home port of Hare Bay aboard a two-masted schooner in early June of 1934. Along her route to St. John's, the crew of the *Ethel Collett* tell each other stories of life and death on the sea. They relive some of Newfoundland's richest historical moments, from shipwrecks and sealing disasters to political strife and financial ruin. But little do they know that they are heading toward one of the most astonishing tales of them all: their own.

ALSO BY GARY COLLINS

"*Soulis Joe's Lost Mine* is a number of stories in one: it's a great mystery-adventure; it's a fascinating look at prospecting for precious metals; and it's a heart-warming story about the importance of family pride."
The Halifax Chronicle Herald

In the summer of 2008, writer Gary Collins teamed up with Allan Keats, a great-grandson of legendary prospector Soulis Joe, and they set out to unearth the secret of Soulis Joe's lost silver mine. After many weeks and months spent combing the island of Newfoundland, Gary Collins figured it out. Come along for the trip and discover it for yourself.

CANADIAN ABORIGINAL BOOKS FOR SCHOOLS 2010–2011 SELECTION

ALSO BY GARY COLLINS

"This tale, set by the sea in Newfoundland, is told in a simple repetitive refrain that will capture the imagination of young readers. . . . Illustrations by Scott Keating . . . capture the beauty of Newfoundland and the many seasons and moods of the ocean."
Atlantic Books Today

Maggie Rose's favourite song is "What Colour's the Ocean Today," a song she made up with her Grandy. Join Maggie Rose and Grandy through Winter, Spring, Summer, and Fall and see what colour the ocean is today! Parents and children will both enjoy reading (and singing) this book over and over again. With colourful illustrations and sheet music in the back, *What Colour is the Ocean?* is sure to become a family favourite . . . no matter what season it is!

WINNER OF THE 2010 ATLANTIC BOOK AWARDS
LILLIAN SHEPHERD MEMORIAL AWARD FOR EXCELLENCE IN ILLUSTRATION